637 Gorillas on the Run and Other Feats, Facts and Astonishing Stats

The Best of Number Crunch: Volume 2

John S. Croucher
Cartoons by Roger Harvey

16
EasyRead Large

RHYW

Copyright Page from the Original Book

Woodslane Press Pty Ltd

7/5 Vuko Place, Warriewood, NSW 2102

Email: info@woodslane.com.au

Website: www.woodslane.com.au

Print edition first published in 2010 by Woodslane Press
Ebook edition published by Woodslane Press in 2011

National Library of Australia Cataloguing-in-Publication entry

Author:	Croucher, John S.
Title:	637 Gorillas on the Run (electronic resource): and other feats, facts and astonishing stats : best of number crunch 2 / John Croucher.
Edition:	1st ed.
ISBN:	9781921683930 (ebook)
Subjects:	Newspapers—Sections, columns, etc. Curiosities and wonders. Mathematical statistics. Statistics. World records.
Dewey Number:	519.5

Design and layout by Jenny Cowan Design

Set in 16 pt. Verdana

TABLE OF CONTENTS

Dedicated to my dear friend

Julian Leslie

PREFACE

It is no secret that many people have a fascination with numbers – especially in this computer age. There is now the opportunity for recording and accessing more information than was ever dreamed possible by our ancestors. The seemingly endless stream of statistics quoted by commentators during sporting broadcasts is one example of our thirst for facts and figures; we digest them and draw our own conclusions on their meanings.

The genesis of this book was the response to my newspaper columns that appear around the world; the public's imagination has been captured by numbers both amazing and challenging. For over ten years the weekly 'Number Crunch' column in the *Sydney Morning Herald* and *The Age* in Melbourne have provided riveting data to Australia, while the weekly 'Statistically Speaking' column that appeared recently in the *Mail on Sunday* in the United Kingdom offered international readers the opportunity to be fascinated by statistics relevant to their own part of the world. These columns are read by millions of readers each week and provide testimony to just how much fun numbers can be.

For me the fascination with figures began very early. I have always been interested in numbers – for cricket, football and tennis especially but indeed for all sports. I pored over almanacs and sporting results. Numbers talked, sang, even shouted at me. I could see that there was so much to be achieved by analysing them. They told you how things were. They gave you a measure – a standard – against which other things, past, present and future, could be assessed.

This obsession has led to me being known as 'Professor FootyTAB' on television in Australia and 'Dr Number from Down Under' in my BBC Radio Quiz in Scotland, to my weekly columns in newspapers, and now to this book.

637 Gorillas on the Run: Best of Number Crunch Volume II builds on *Eighteen Months on the Toilet: Best of Number Crunch Volume I* and contains a new collection of facts to entertain and inform. It also contains a collection of some of the more captivating statistics I have found. It is easy to dip into them, staring at any section. With over 2,500 facts and figures and around 150 matching humorous quotations, there is plenty of information to keep the reader absorbed and amused. While there are specific categories, the 'Quirky' section in particular contains those that have established themselves as the most popular over the years.

The sources of these statistics are many and varied – ranging from encyclopaedias, books of facts, news reports, surveys and journal articles – all current at the time of writing. Those items that reveal the views and attributes of populations rely on survey data and are of course dependent on the particular respondents in the sample. But they are all from reliable samples of reasonable size and are often the figures that promote the most discussion among readers. A 'Further Reading' section is provided for those wishing to learn more.

As was the case in volume I, this book would not have been possible without the loving assistance of my wonderful wife Rosalind, without whose laughter, patience and guidance made it such an enjoyable project. She is still, and always will be, the light of my life.

JOHN S. CROUCHER
2010

QUIRKY

Always get married early in the morning. That way, if it doesn't work out, you haven't wasted a whole day.

Mickey Rooney, US actor

Half the lies they tell about me aren't true.

Yogi Berra, former US baseball player and manager

***

Proportion of surveyed people in the UK who say they have fallen in love in a supermarket:
2 per cent

Most pancake tosses in two minutes (Ralf Laue, Germany):
399
year: **1997**

Life expectancy of an aardvark in activity:
up to 24 years
weight: **40–65 kilograms**
number of teeth (located near the back of the jaw): **20**

Amount of beer sold in the world each year:
about 135 billion litres

Number of canals in Venice: **150**
total length of the Grand Canal:
3 kilometres

Year that author Enid Blyton created Noddy:
1949

Number of rounds in the longest boxing match on record (between Andy Bowen and Jack Burke in New Orleans):
110
time taken: **7 hours 19 minutes**

Number of times the word 'cabbage' appears in the works of Shakespeare (in Act 1, scene 1 of *The Merry Wives of Windsor*):
1

Estimated proportion of the world's population that eat with chopsticks: **30 per cent**
reported ideal length of a chopstick:

24 centimetres

Average number of microscopic saliva droplets sprayed in one minute of talking:
about 300

Nobody believes the official spokesman, but everybody trusts an unidentified source.

Ron Nesen, former White House press secretary

***

Number of pieces into which a 28 centimetre cucumber was sliced (by Dr. S. Ramesh Babu): **120,060**
time taken: **2 hours 52 minutes 21 seconds**
year: **1998**

Ranking of 'the' in the most written words in the English language: **1**
'of': **2**
'and': **3**

Proportion of the average fart that is:
nitrogen: **20–90 per cent**
hydrogen: **0–50 per cent**
carbon dioxide: **10–30 per cent**
methane: **0–10 per cent**
oxygen: **0–10 per cent**
other: **21 per cent**

Weight of garlic the average American eats each year: **about 1.3 kilograms**
number of times the word 'garlic' appears in the entire works of Shakespeare: **4**

Highest score in a round of golf at the US masters (by Charles Kunkle): **95 (49 and 46)**
number of strokes over par: **23**
year: **1995**

Number of rainy days per year on Mount Waialeale, Hawaii:
up to 350
average annual rainfall:
11,440 millimetres

Proportion of the people hanged in Britain in 1785 who were under the age of 20:
90 per cent

Number of teeth on a hedgehog: **36**
number of spines: **5,000**

Height of an iceberg classified as 'very large' above water:
at least 75 metres

Number of species of mistletoe:
about 2,000

Proportion of surveyed Britons who say they have had oral sex while driving:
35 per cent

Number of horns on the horned owl: **0**

Number of countries with a national flag that is not the same on both sides:
1 (Paraguay)

Number of US presidents to have regained the presidency after being voted out of office:
1 (Grover Cleveland)

I never drink anything stronger than gin before breakfast.

W.C. Fields, US comedian and actor

***

Number of dogs in Paris:
about 200,000
reported weight of excrement left by dogs on Paris streets and footpaths each day:
10–15 tonnes

Weight of snails eaten each year in France:
40,000 tonnes
in Portugal: **4,000 tonnes**

Average number of seeds in a strawberry:
about 200

Number of American presidents who have appeared in a shirt commercial:
1 (Ronald Reagan)

Year that the Venetian gondoliers became subject to breathalyser tests:
2005

Estimated proportion of the population of Wales who speak Welsh:

about 22 per cent

Number of years spent in prison by Marie-Augustin, Marquis de Pelier of Brittany, spent in a French prison for whistling at Marie Antoinette as she took her seat in a Paris theatre:
50 years
his age at the time: **22**

The speed limit for a steam-driven car in the UK in 1864:
2 miles per hour (3.2 kilometres per hour)

Number of times the word f**k appears in the movie
Pulp Fiction: **155**
in *Reservoir Dogs:* **177**

Estimated total weight of all people on Earth:
380 million tonnes

Number of MMS messages sent in Australia on mobile phones each year:
about 50 million

Proportion of female swimwear sold in France that are bikinis:
50 per cent

Proportion of surveyed Australians aged less than 40 years who expect their parents to fritter away their inheritance:
50 per cent

Year a getaway car was used for the first time in a robbery (in Paris): **1901**

Once you've put one of his books down, you simply can't pick it up again.

Mark Twain, US author and humourist (talking about Henry James)

***

Reported number of shopping trolleys used by Woolworths, Safeway and Big W stores in Australia in 2010:
around 300,000
number of stores: **over 900**
proportion of trolleys that are replaced each year:
about 5 per cent

Number of rivets in the Eiffel Tower:
about 2.5 million
Sydney Harbour Bridge:
about 6 million

Number of plagues of Egypt: **10**

Number of men involved in the Charge of the Light Brigade, Crimean War: **673**
number killed: **118**
wounded: **127**
year: **1854**

Estimated number of Americans who have obtained degrees from "non-accredited" institutions:
100,000

Estimated number of Australians living overseas:
about 1 million
surveyed proportion of these who "still call Australia home":
63 per cent

Proportion of surveyed Christmas shoppers in Australia who did not know what they were going to buy when they left home:
50 per cent

Most squat thrusts in one hour (by Paddy Doyle, UK):
3,743
year: **1998**

Population of California, USA in 2010:
about 37 million
as a proportion of all US residents:
1 in 8

Proportion of surveyed people worldwide who say they scrunch toilet paper into a ball: **42 per cent**
fold it over: **42 per cent**
wrap it around their hand: **11 per cent**
other: **5 per cent**

Largest gathering of people dressed as gorillas (at the *Great Gorilla Fun Run* in London): **637**
year: **2005**

Number of cows in a flink:
12 or more

The 50-50-90 rule: Anytime you have a 50-50 chance of getting something right, there's a 90% probability you'll get it wrong.

Andy Rooney, US radio and television writer

***

Proportion of customers who turn right when entering a supermarket when there is a choice of direction: **75 per cent**

Number of patents lodged with the US Patent and Trademark Office that are designed to cure snoring: **over 300**

Proportion of surveyed Australians who make New Year resolutions: **43 per cent**
proportion of these that involve exercise: **75 per cent**
a better work-life balance: **42 per cent**
going on a diet: **39 per cent**
quitting smoking: **17 per cent**
reducing alcohol consumption: **10 per cent**

Proportion of surveyed people worldwide who install toilet paper rolls over the top: **72 cent**
around the back: **28 per cent**

Usual range of length of tapeworms found in a human: **5 millimetres to 15 metres** longest on record: **33 metres**

Proportion of schoolteachers in Australia who are female:
68 per cent
in 1985: **58 per cent**

Proportion of surveyed Australian men who own a suit they no longer want:
87 per cent

Proportion of Australian families who live in rental accommodation: **1 in 3**

Proportion of surveyed Americans who say that the most frustrating sound to hear is a:
car alarm: **39 per cent**
jackhammer: **28 per cent**
baby crying: **21 per cent**
dog barking: **13 per cent**

Proportion of surveyed Australian men who do not usually wash their hands after using a public toilet:
1 in 3
women: **1 in 10**

Dimensions of the world's largest cardboard box (in Helmond, Netherlands):
10.04 metresx4.02 metresx2.20 metres
year made: **2007**

Whenever I'm caught between two evils, I take the one I've never tried.

Mae West, US actress

***

Number of streets in the USA: **about 5.5 million**
number that are named:
Peppermint: **95**
Chocolate: **27**
Caramel: **26**
Licorice: **12**
Vanilla: **11**

Proportion of surveyed Australians who say they average less than 6.5 hours of sleep a night:
20 per cent
who say they feel chronically sleepy during the day:
11.7 per cent

Greatest recorded snowfall in the USA in a 24-hour period (in Silver Lake, Colorado):
195.6 centimetres
date: **14–15 April 1921**

Date on which the population of the USA reached 300 million:
about 17 October 2006

year it is estimated to reach 400 million:
2043

Proportion of surveyed Australians who regularly take sleeping tablets: **10 per cent**
proportion of these who do so every night: **1 in 3**

Amount paid at auction in October 2007 for a lock of Ernesto "Che" Guevara's hair and related items:
about $A131,800
length of the lock:
about 8 centimetres

Estimated number of foreign domestic helpers in Hong Kong:
about 242,000
proportion of these from:
the Philippines: **53 per cent**
Indonesia: **43 per cent**

The age with the greatest number of people in Australia at 30 June 2009: **38 years**
number of people at this age:
335,100

Dimensions of the world's largest ruby (in Beijing, China):
30mmx138mmx145mm
weight: **8184 grams**
year: **2005**

Number of Australians who have been awarded the Victoria Cross: **96**
number of these in World War I: **64**
World War II: **20**

Everything is funny as long as it is happening to somebody else.

Will Rogers, US cowboy and comedian

***

Proportion of surveyed English adults who do no exercise at all:
50 per cent
less than 30 minutes each week: **30 per cent**
at least 30 minutes: **20 per cent**

Reported number of plastic bottles discarded by Americans each hour:
about 2.5 million

Factor by which you are less likely to laugh if you are home alone than if someone is home with you:
30 times

Average length of an attack of the hiccups:
about 5 minutes
average time an adult hiccups in a lifetime:
about 2,300

Average weight of personal items British women carry in their handbags:
2.4 kilograms

Recommended length of a 'power nap' taken in the afternoon:
20 to 30 minutes

Proportion of surveyed Australians who say they have no religion: **30 per cent**
in 2001: **27 per cent**

Estimated number of homeless people in Australia on any given night: **about 100,000**
proportion who are women: **about 40 per cent**
proportion who are aged under 25: **about 50 per cent**

Average length of a sleepwalking episode:
about 6 minutes

Proportion of surveyed Australian adults who could not correctly spell the word:
"questionnaire": **40 per cent**
"definitely": **37 per cent**
"liaison": **31 per cent**
"occurred": **27 per cent**

Average sheets of toilet paper used by Americans each day:
about 57
average weight used per year:
23 kilograms

Proportion of the UK population that live in:
England: **83.8 per cent**
Scotland **8.4 per cent**
Wales: **4.9 per cent**
Northern Ireland: **2.9 per cent**

When I was a kid I got no respect. I told my old man, "I'm sick and tired of running around in circles." He got mad. He nailed down my other foot.

Rodney Dangerfield, US comedian and actor

***

Proportion of surveyed Australians who say they have
no religion:
about 19 per cent

Rate at which the world's population is currently growing each year:
1.2 per cent

Proportion of surveyed Australians who feel that modern society is unsympathetic to disadvantaged people:
45 per cent

Proportion of surveyed US males for whom the oldest underwear in their drawer is less than one year old:
17 per cent
1–4 years old: **58 per cent**
5–9 years old: **15 per cent**
at least 10 years old: **11 per cent**

Proportion of surveyed Americans who believe that their standard of living will be better in five years time:
62 per cent

Proportion of surveyed Australians aged 15 and over who meet the minimum recommendation of exercising three times a week for 30 minutes:
42.8 per cent

Proportion of surveyed Australians who feel they could recycle more or buy more products with less packaging:
54 per cent

Number of registered voters in the 2008 US Presidential election: **169 million**
number who actually voted: **131.3 million**
average age of workers at the polling booths: **72**

Number of islands (excluding external territories) that belong to Australia: **8,222**

Number of Australians aged over 100 years as at 30 June 2009:
about 3,700
number of these who were women: **2,900**
forecast total number aged over 100 years by 2055:
about 78,000

Proportion of surveyed Australian women aged 18–64 who say buying gifts for others makes them feel happy:
36 per cent
men: **19 per cent**

What if nothing exists and we're all in somebody's dream? Or what's worse, what if only that fat guy in the third row exists?

Woody Allen, screenwriter, film director, actor, comedian, writer and musician,

***

Estimated number of people in the USA in 2009 who experience homelessness at least once a year:
up to 3.5 million

Proportion of surveyed Americans who said they lie about reading classical novels just to impress friends:
65 per cent

Number of US Presidents who have been an only child (not counting half brothers and sisters as siblings):
0

Proportion of surveyed Americans aged 25–54 who, when having a shower, say they sometimes think about their:
‘to do’ list: **56 per cent**
problems and worries: **41 per cent**
work: **28 per cent**
weight: **20 per cent**
fitness: **14 per cent**

who daydream: **33 per cent**

Proportion of surveyed Australians who are satisfied with life in their country:
82 per cent

The most recent year that can be written upside-down and right side-up and appear the same: **1961**
the next such year: **6009**

Diameter of the wire in a standard paper clip:
1 millimetre

Number of men for every 100 women in Australia:
99
at age 75: **69**

Proportion of surveyed Australians who regularly receive a Christmas gift they don't like:
72 per cent
average number of unwanted gifts received each Christmas:
1.5

Proportion of surveyed churchgoing Australians who read the Bible:
every day: **21 per cent**
a few times a week: **14 per cent**
once a week: **6 per cent**
occasionally: **24 per cent**

hardly ever or never: **35 per cent**

Number of Prime Ministers of Australia since Federation in 1901: **27**

Greatest daily temperature range recorded in Australia (at Eyre, WA): **37.40C**

Some people talk in their sleep. Lecturers talk while other people sleep.

Albert Camus, French Algerian author, philosopher and journalist

***

Proportion of surveyed women who say they have 1–3 disturbed night's sleep each week:
66 per cent
men: **50 per cent**

Number of stars on the American flag: **50**
stripes: **13**

Number of flags with just one colour (green): **1 (Libya)**

Reported number of people worldwide who die from vending machines falling on them:
about 13

Most common score given by surveyed Australian men on a scale of 1–10 when asked to rate their satisfaction with their life: **8**

Proportion of surveyed Americans in 2009 who say they have given an unwanted present to someone else as a gift:
36 per cent
in 2008: **24 per cent**

Proportion of surveyed Americans who find the most annoying thing in a conversation is the word or phrase:
'whatever': **47 per cent**
'you know': **25 per cent**
'it is what it is': **11 per cent**
'anyway': **7 per cent**
'at the end of the day': **2 per cent**

Proportion of surveyed habit-free smokers who were able to quit after the first attempt: **40 per cent**
2–3 attempts: **38 per cent**
4 or more attempts: **22 per cent**

Proportion of surveyed Americans who say they flush public toilets:
with their shoe: **41 per cent**
normally: **30 per cent**
using a paper towel: **17 per cent**
other: **12 per cent**

Experience is that marvellous thing that enables you to recognize a mistake when you make it again.

Franklin P. Jones

***

Estimated number of words in the English language: **about 1 million**
Chinese (various dialects): **over 500,000**
Japanese: **232,000**
Spanish: **over 225,000**
Russian: **195,000**

Proportion of serious injuries in Australia from surveyed ladder falls that occurred while:
undertaking home improvement: **62 per cent**
pruning trees/hedges: **13 per cent**
cleaning gutters/roof: **11 per cent**
picking fruit: **6 per cent**
other: **8 per cent**

Proportion of surveyed American believers in God in 2010 that say God is a man: **38 per cent**
a woman: **1 per cent**
both a man and a woman: **11 per cent**
neither a man or a woman: **34 per cent**

Proportion of surveyed Australians in 2010 who thought Australia had changed for the worse in the past 20 years:
52 per cent
for the better: **39 per cent**
no change: **9 per cent**

Proportion of hotel guests in Australia who flood the bathroom that are women: **72 per cent**
who leave their room in a shambles:
69 per cent

Proportion of hotel guests in Australia caught naked outside their room who are women:
29 per cent

Reported number of letters and emails addressed to President Obama received by the White House each day in 2010:
about 20,000

Energy used per hour by fidgeting:
about 61 kilojoules
Longest distance walked with a milk bottle balanced on the head (Ashrita Furman, USA): **130. 3 kilometres**
time taken: **23 hours 35 minutes**
year: **1998**

Most handshakes with different people by an individual in eight hours (by Yogesh Sharma, India): **31,118**
year: **1996**

Average pairs of shoes surveyed British women purchase each year: **4**
proportion of shoes she wears only once: **25 per cent**

Proportion of surveyed Americans who believe in God: **82 per cent**

You know why fish are so thin? They eat other fish.

Jerry Seinfeld, US comedian and actor

***

Most chin-ups in one hour (Stephen Hyland, UK): **908**
year: **2009**

Most push-ups in one hour using one arm and the back of the hand (Doug Pruden, Canada): **1,025**
year: **2008**

Proportion of surveyed US women who say they feel better after crying: **85 per cent**
of men: **73 per cent**

Time the average Australian male spends shaving in his lifetime:
about 3,000 hours

Proportion of surveyed British women who remember the first pair of shoes they bought with their own money:
92 per cent
who felt remorse at throwing away a pair of old shoes:
96 per cent

Estimated proportion of Australians who are sleep deprived:
63 per cent

Number of items posted in a typical December in Australia:
460 million

Proportion of surveyed Australians who do 30 minutes of exercise each day:
33 per cent

Longest female dreadlocks (Asha Mandela, USA):
2.59 metres
year: **2008**

Heaviest vehicle pulled by the eyelids (Dong Changsheng, China):
1500 kilograms
distance: **10 metres**
year: **2006**

Proportion of surveyed people who say they can sense when they are being watched:
87 per cent

Proportion of surveyed Australians who feel it is acceptable for politicians to stretch the truth in the heat of a verbal exchange:
23 per cent

Number of years late a book *(The Law of Nations)* borrowed from New York Society by George Washington was returned late: **221**
date it was borrowed: **5 October 1789**
estimated overdue fee: **about $US300,000**

Estimated proportion of Australians born this century who will reach age 100:
50 per cent

If you are first you are first. If you are second you are nothing.

Bill Shankly, former British football manager

***

Proportion of surveyed Australian women in May 2010 aged between 45 and 56 who said they were very satisfied with their life: **31 per cent**
in May 2009: **45 per cent**

Time it took the first swimmer to cross the English Channel (Captain Matthew Webb):
21 hours 45 minutes
year: **1875**
current record (Petar Stoychev, Bulgaria):
6 hours 57 minutes 50 seconds
year: **2007**

Weight of nails, coins and scrap metal removed from a man's stomach in Peru in November 2009:
900 grams

Proportion of surveyed Americans who say the day they dread the most is:
a visit to the dentist: **33 per cent**
Mondays: **16 per cent**
the day their tax is due: **15 per cent**

Proportion of Australians who agree that it is necessary to wash your hands with soap after going to the toilet:

98 per cent

SPORT

He wears a number 10 jersey. I thought it was his position, but it turns out it's his IQ.

George Best, British football player

I have a tip that can take 5 strokes off anyone's golf game. It's called an eraser.

Arnold Palmer, US golfer

***

Winning time of the men's 4x200 metre freestyle swimming relay at the 1924 Paris Olympics (a world record):
9 minutes 53.4 seconds
current men's world record: **7 minutes 4.6 seconds**
current women's world record: **7 minutes 53.42 seconds**

Number of licensed bookmakers in Australia:
about 610

Number of Test matches played by cricketers:
Sir Garfield Sobers: **93**
Geoffrey Boycott: **108**
Sunil Gavaskar: **125**
Allan Border: **156**

Year that yellow tennis balls replaced white ones at Wimbledon:
1986
number of tennis balls used during the current Championships:
about 42,000

Shane Warne's top score in First Class cricket (for Hampshire, England, 2005): **107**
his top Test score (versus New Zealand, Perth, 2001–2): **99**

Number of times the modern Olympic Games have been held in London:
2 (1908 and 1948)

Most gold medals won by Australia at a single Olympics:
17 (Athens, 2004)
most silver medals: **25 (Sydney, 2000)**
most bronze medals: **23 (Atlanta, 1996)**

Year the first Olympic Games took place: **776B.C.**
number of events: **1 (a footrace)**

Most points scored in the opening game of an NBA season in the USA (by Michael Jordan, Chicago Bulls): **54**
year: **1989**

Height of a badminton court net at the middle: **1.524 metres**
a table tennis net: **15.24 centimetres**

Estimated proportion of Australian adults who attend at least one sports event a year:
48 per cent

Length of the world's largest swimming pool (in Algarrobo, Chile): **1,013 metres**
area covered: **about 8 hectares**
year completed: **2006**

I know I'm getting better at golf because I'm hitting fewer spectators.

Gerald R. Ford, former US President

***

Longest time heading a regulation soccer ball without dropping it (by Thomas Lundman, Sweden):
8 hours 32 minutes 3 seconds
year: **2004**

Number of cricket teams that have played five or more Tests in a calendar year and won them all:
2 (Australia, 8 Tests, 2000; Australia, 10 Tests, 2006)

Number of players who have scored three double centuries in Test cricket in a calendar year:
2 (Don Bradman and Ricky Ponting)

Number of 'Wisden's Cricketers of the Century':
5 (Bradman, Sobers, Hobbs, Warne, Richards)
number who were knighted: **4 (Shane Warne was not)**

Length of the longest jump by a motorcross rider (Robbie Maddison, Australia):
98.15 metres
Year: **2008**

Number of players in a netball team: **7**

World record for underwater breath holding (David Blaine, US):
17 minutes 4.4 seconds
year: **2008**

Most wickets to fall in a single day's play of a Test match (England v Australia at Lords): **27**
number of runs scored in the day: **157**
year: **1888**

Fewest runs scored in a full day's play in a cricket Test (Pakistan v Australia): **95**
wickets that fell: **12**
year: **1956**

Proportion of surveyed UK football fans who would miss a family friend's funeral to attend a match:
25 per cent

Most runs scored in a single day's play in a cricket Test
(England v India, 2nd day at Old Trafford):
588 (England went from 2/173 to 8/571, India 0 for 190 at stumps)
wickets that fell: **6**
year: **1936**

Highest opening partnership in Test cricket (by Graeme Smith and Neil McKenzie, for South Africa v Bangladesh): **415**
year: **2008**

I had a wonderful experience on the golf course today. I had a hole in nothing. Missed the ball and sank the divot.

Don Adams, US actor and comedian

***

Number of hoops on a croquet court: **6**

Starting date of the first cricket Test Match (England v Australia):
15 March 1877
the first Wimbledon tennis tournament:
9 July 1877

Width of the front of the home plate in baseball:

43.2 centimetres (17 inches)

Highest percentage of a team's completed innings in Test cricket made by a single batsman (Charles Bannerman for Australia in Melbourne, during the first ever Test match):
67.3 per cent

Size of the crowd at Caulfield Racecourse Melbourne in October 1946 to see *Bernborough* run in the Caulfield Cup:
108,213
weight carried: **68 kilograms**
finishing position: **5**

Number of rugby league premierships won by South Sydney (the highest):
20

Minimum number of balls faced to score a century in Test cricket (Viv Richards, West Indies v England, 1985–6):
56

Number of strokes it took American golfer Floyd Rood to hit a golf ball across the USA: **114,737**
time taken: **13 months**
number of balls lost: **3,511**
years: **1963–4**

Reported number of golf balls left on the moon:
2 or 3

Golf is a fascinating game. It has taken me nearly forty years to discover that I can't play it.

Ted Ray, British golfer

***

Number of Test matches played by English cricketer W.G. Grace:
22
total runs scored: **1,098**
batting average: **32.29**
highest score: **170**

Australian cricketer Don Bradman's Test batting average before his last innings duck:
101.39

Proportion of surveyed Australian women who say they do no sport or exercise because they are not interested:
19 per cent

Official number of players in a:
field hockey team: **11**
volleyball team: **6**
basketball team: **5**

Most points scored by an individual NBA finals game (Elgin Baylor, Lakers v Celtics): **61**

year: **1962**

Number of Grand Slam singles titles won by Venus Williams:
7 (2xUS Open, 5xWimbledon)
Serena Williams:
**13 (3xUS Open, 5xAustralian Open, 1xFrench Open,
4xWimbledon)**

Number of golf courses in:
New Zealand: **about 420**
Scotland: **about 540**
Australia: **about 2,500**

Distance of the Sydney to Hobart yacht race:
628 nautical miles (about 1164 kilometres)
current race record (by Wild Oats XI):
1 day 18 hours 40 minutes 10 seconds
year: **2005**

Proportion of surveyed Australian adults who have never been to a gym: **1 in 3**

On a par 5 hole in golf, number of strokes for a double eagle (albatross): **2**

Highest number of Grand Slam single tennis titles won (by Australian Margaret Court, nee Smith): **24**

You wouldn't have won if we'd beaten you.

Yogi Berra, former US baseball player and manager

***

Number of people aged 15 and over in Australia in 2007 who were involved in instructing, coaching or teaching sport:
659,000
refereeing or umpiring:
381,000

Number of laps drivers complete during the Indianapolis 500:
200
length of the race:
804 kilometres (500 miles)

Proportion of Australian men who dream of competing at an international sporting event: **45 per cent**
of women: **21 per cent**

Longest distance jumped on a Harley-Davidson (Seth Enslow from the US, in Sydney):
55.99 metres
year: **2010**

Most goals scored by a team in the National Hockey League in a single match (by the Montreal Canadiens over the Quebec Bulldogs): **16**
year: **1920**

Amount of oxygen consumed by an athlete while running 100 metres:
about 7 litres

Number of footballs an NFL team uses in a single season:
about 1,000

Most NASCAR road-course wins (Jeff Gordon): **9**
year of his last win: **2006**

Lowest score in the Women's British Open golf (Karrie Webb in 1997, Karen Stupples in 2004):
269

Longest distance on a treadmill (Mike Bus, UK):
832.4 kilometres
time it took: **7 days**
year: **2010**

Number of lost golf balls reportedly collected by a man who dived by night into water hazards at South Korean golf courses:
26,000

Year that numbered shirts were first used to identify players in a British FA Cup final: **1933**

A computer once beat me at chess, but it was no match for me at kick boxing.

Emo Phillps, US entertainer and comedian

***

Number of official no-hitters in US Major League Baseball between 1876 and 2009: **263**
number due to pitcher Nolan Ryan: **7 (highest)**

Lowest total score in the British Open golf championship (Greg Norman at Royal St George): **267**
year: **1993**

Number of cricket Tests in which Don Bradman was captain:
24
number of these won by Australia: **15**

Number of cricket Tests in which Allan Border was captain: **93**
number of these won by Australia: **32**

Number of cricket Tests in which Steve Waugh was captain: **57**
number of these won by Australia: **41**

Age of the youngest winner of the Women's French Tennis Open (Monica Seles, Yugoslavia):
16 years 6 months
year: **1990**

Age of the oldest player (Julio Franco for the New York Mets) to hit a home run in Major League Baseball in the US:
48 years 254 days
year: **2007**

Age of the youngest winner of the Men's French Tennis Open:
(Michael Chang, USA):
17 years 3 months
year: **1989**

Sports betting turnover in Australia in 2009-10:
$2.9 billion

Reported decibel level of the grunting of Russian tennis player

Maria Sharapova: **about 103**
decibel level of a lion's roar: **110**

There is a fine line between fishing and just standing on the shore like an idiot.

Steven Wright, US actor and writer

***

Year that Indian cricketer Sunil Gavaskar became the first batsman to score 10,000 runs in Test cricket: **1987**

The number between 17 and 19 on a dartboard: **3**

Width of a singles tennis court: **8.23 metres**
doubles: **10.97 metres**

Number of consecutive strikes in bowling for a turkey: **3**

Number of Olympics at which Australia has won no medals:
1 (St Louis, 1904)

SEX

You know that look women get when they want sex? Me neither!

Drew Carey, US comedian and actor

I tell you, my wife, she likes to talk during sex. The other night she called me from a motel.

Rodney Dangerfield, US comedian and actor

***

Proportion of surveyed women who said they have achieved orgasm by stimulation of their breasts only:
29 per cent

Proportion of surveyed British women who fantasise about having sex:
while being tied up: **28 per cent**
with two or more men: **28 per cent**

Proportion of surveyed Australian men aged 16–60 who had cheated on their partner in the past year:
5 per cent
women: **3 per cent**

Proportion of surveyed French men aged over 40 who had a willingness to improve their sexual performance:
60 per cent
of British men:
46 per cent

Proportion of surveyed Australian men aged 16–60 who said they had more than one sexual partner in the past year:
15 per cent
women: **8 per cent**

Proportion of surveyed women who say they regularly fake orgasms:
52 per cent

Proportion of surveyed British women who would give up sex if it meant they would live to age 100:
48 per cent
of British men:
31 per cent

Proportion of surveyed Australian men aged 25–45 who would like to sleep with someone new as soon as possible:
23 per cent
who think a cosy night is a prelude to sex:
62 per cent

Proportion of surveyed Britons aged 65 and over who wished they had more sex when they were younger:
70 per cent
travelled more: **57 per cent**
married someone else: **21 per cent**

Proportion of surveyed Australian women aged 16–25 who have had sex at least once without a condom:
77 per cent
for whom the pill is the most trusted form of contraception:
63 per cent

Sex is nature, and I believe in going along with nature.

Marilyn Monroe, US actress

***

Estimated number of abortions in Australia that are to women:
aged below 20: **18 per cent**
aged 20 to 30: **49 per cent**
above 30: **33 per cent**

Proportion of surveyed women in the USA who, after having met someone on the internet, had sex on the first date:
1 in 3
proportion of these who did not use a condom:
75 per cent

Amount a Western Australian barmaid was fined in October 2007 for showing patrons she could crush beer cans between her exposed breasts: **$1,000**
her age: **31**
amount which her female co-worker was fined for assisting her to hang spoons from her nipples: **$500**

Proportion of surveyed single Britons who are unhappy with their sex life:

63 per cent
married Britons:
31 per cent

Proportion of surveyed British men who fantasise about having sex with:
someone else's partner: **58 per cent**
a work colleague: **53 per cent**
a stranger: **42 per cent**
a friend's partner: **36 per cent**

Average number of sex partners had by Hong Kong males in their lifetime: **14**
females: **4**

Proportion of surveyed Britons who say they:
never have sex: **18 per cent**
have sex every day: **2 per cent**
more than once a day: **1 per cent**

Proportion of surveyed Americans who think it is very important:
to have fidelity in a marriage: **93 per cent**
to have a happy sexual relationship: **70 per cent**

Proportion of surveyed Australian women who had an orgasm at their last heterosexual encounter: **31 per cent**
those involved in lesbian sex: **24 per cent**

Proportion of surveyed German men with unsatisfying sex lives who said they use hard work as a diversion: **38 per cent**
women: **35 per cent**

Sex is a two-way treat.

Franklin P. Jones

***

Proportion of surveyed executives with a job involving regular travel who say it is impossible to have a satisfying sex life:
45 per cent

Proportion of surveyed American adult men who have had two or more sexual partners in the past year:
17 per cent
women: **10 per cent**

Proportion of surveyed Australian adults who say they have no sex life at all:
15 per cent

Proportion of surveyed American adults aged 20–59 who say they have had sex at least once:
96 per cent
who lost their virginity before age 15:
16 per cent

Proportion of surveyed Australian women who say they prefer regular intimate moments over sex:
80 per cent
of men: **61 per cent**

Proportion of surveyed people worldwide who feel that the sexiest profession is to be a:
model: **21 per cent (highest)**
nurse: **10 per cent**
doctor: **10 per cent**
masseur: **10 per cent**

Proportion of surveyed British women who said they fantasise about:
fire-fighters: **54 per cent**
soldiers: **28 per cent**
businessmen: **27 per cent**
doctors: **26 per cent**
footballers: **20 per cent**
milkmen: **0.8 per cent**

It isn't premarital sex if you have no intention of getting married.

Drew Carey, US comedian

***

Proportion of surveyed Australian women who prefer to have sex in the dark due to a poor body image:
60 per cent
who say their excess weight sometimes makes them disinterested in sex altogether: **69 per cent**
who say their weight makes them embarrassed to be seen or touched: **73 per cent**

Estimated amount spent on prostitutes and strippers in Australia in 2007: **$1.13 billion**

Proportion of surveyed British women who had slept with more than 10 partners in the first 5 years since losing their virginity:
25 per cent
of men: **20 per cent**

Proportion of surveyed British women who would rather go shopping than have sex:
40 per cent
who, if married more than 20 years, would be happy with no sex life at all:
50 per cent

who describe themselves as happily married:
94 per cent

Estimated proportion of younger Australians who have tried oral sex before the age of 30:
90 per cent

Average number of sexual partners of surveyed British women by the age of 21: **9**
men: **7**

Proportion of surveyed teenagers in the US who have sent nude or partially-clothed images of themselves to someone by email or mobile phone: **20 per cent**
sent sexually suggestive electronic messages: **40 per cent**

Proportion of surveyed Australian women aged 25–45 who say that rushed foreplay is the biggest issue in sex for them:
38 per cent

Proportion of surveyed Australian men aged 25–45 who complain that women never initiate sex:
35 per cent

Proportion of surveyed people in South Australia who say that back pain is slowing down their sex life:
30 per cent (highest)
in Tasmania: **23 per cent**

in Queensland: **13 per cent (lowest)**

When I came here, I couldn't speak a word of English, but my sex life was perfect. Now my English is perfect but my sex life is rubbish.

Julio Iglesias, Spanish singer

***

Proportion of surveyed Chinese women with male partners who said they had orgasms during sex:
always: **8 per cent**
often: **26 per cent**
sometimes: **50 per cent**
rarely or never: **16 per cent**

Proportion of surveyed women in the US who would rather go without sex for two weeks than go without the internet for two weeks: **46 per cent**
men: **30 per cent**

Proportion of surveyed British women who believe the way they dress affects their day at work:
78 per cent
who would wear a skimpy outfit to work to get a bonus or promotion:
27 per cent

Proportion of surveyed Australian baby boomers who would be happy to have sex outdoors: **21 per cent**
Gen X and Gen Y: **10 per cent**

Proportion of surveyed people who own a sex toy:
in Britain: **43 per cent**
in the US: **43 per cent**
in Sweden: **40 per cent**
in France: **14 per cent**

Proportion of surveyed Australian women who since the age of 15 have experienced unwanted sexual touching:
25.1 per cent
of men: **9.9 per cent**

Average number of times surveyed Greeks have sex each year (world's highest): **164**
Brazilians: **145**
New Zealanders: **122**
Australians: **106**
British: **92**
Japanese: **48**

Proportion of British women living in a same-sex civil relationship who were previously married to a man:
25 per cent

Reported optimal time to spend on sexual intercourse (not including foreplay): **3–13 minutes**
minimum acceptable time: **2 minutes**

Humans are the only animal who can have sex over the phone.

David Letterman, US television host

***

Average age of surveyed sex workers in Australia: **29 years**
proportion who have children: **40 per cent**
who are married or in a de facto relationship: **50 per cent**
who are studying at university or college: **10 per cent**
average annual income: **$66,000**

Proportion of surveyed Australian men aged 20 or older who suffer from erectile dysfunction: **25.1 per cent**
proportion of these who have received any treatment for it: **14.1 per cent**

Proportion of surveyed British women who have sex once or twice a week: **37 per cent**
who say they are unhappy with the amount of sex they have: **34 per cent**

Reported number of sex shops in Queensland in September 2008: **about 180**

Average age of surveyed female sex workers aged 18–57 in Queensland: **32**
proportion who have completed a Bachelor Degree or higher:
20 per cent

Proportion of surveyed British men who had a positive feeling after a one night stand:
80 per cent
women: **54 per cent**

Proportion of surveyed people worldwide who say that sex is fun, enjoyable and a vital part of life:
60 per cent

Proportion of surveyed people worldwide who are happy with their sex life: **44 per cent**
Nigerians: **67 per cent (highest)**
Japanese: **15 per cent (lowest)**

Estimated proportion of de facto relationships in Australia that are same sex:
4 per cent

Number of sex workers in Australia:
about 20,000

estimated number of visits made annually to them: **12–16 million**
average number of clients seen each week by a sex worker: **20**
average charge per visit: **$100**

Proportion of surveyed Australians who say their excess weight seriously impacts on their sex life: **52 per cent**
proportion of women who have tried to lose weight: **81 per cent**
men: **63 per cent**

Kids. They're not easy. But there has to be some penalty for sex.

Bill Maher, US comedian, actor, television host and author

***

Number of erections a male reportedly has per night while sleeping: **4 or 5**
length of each one if undisturbed:
about 10 minutes

Proportion of surveyed Australian men who sometimes look at pornography:
74 per cent

Number of penises from 92 species of animals on display at
the Icelandic Phallological Museum in early 2010: **272**

Proportion of surveyed women in Australia who watch adult movies when staying in hotels:
12 per cent

Proportion of surveyed British men aged 75–85-years-old who say they are sexually active:
38.9 per cent
women: **16.8 per cent**

Proportion of surveyed British men aged 75–85 years-old who say they are still interested in sex: **41.2 per cent**
women: **11.4 per cent**

Average length of an adult kiss:
about 4.5 seconds
amount of energy burnt during a 60 second kiss:
109 kilojoules

Number of sexual partners had so far in a survey of Cambridge University students studying:
medicine: **8 (highest)**
political science: **7**
history: **6**
languages: **5**
theology: **2 (lowest)**

Number of men in Australia in 2008 diagnosed with a sexually transmitted infection (STI): **24,000**
women: **35,000**

Estimated number of sex workers in Thailand: **up to 2.8 million**
number of these who are women: **about 2 million**
estimated proportion of tourist dollars spent there on the sex trade: **10 per cent**

Proportion of surveyed US couples who regularly sleep in separate beds:

23 per cent

Proportion of German men reporting having no sex life who voluntarily take on more responsibility at work:
45 per cent
women: **46 per cent**

I can't believe it. Both my boyfriends are cheating on me.

Lucy Wilde

***

Proportion of surveyed Australian women aged 25–45 who say they want sex every day:
33 per cent
who actually have sex once a month or less:
20 per cent

Proportion of surveyed Australian men aged 25–45 who say they want sex every day:
40 per cent

Proportion of surveyed Australian women aged 25–45 who say they are turned on by dancing:
66 per cent

Estimated proportion of 14 to 17-year-olds in Britain who have looked at pornographic images on the internet: **90 per cent**
who do so more than once per week: **20 per cent**

Proportion of surveyed Australians aged 18-40 who have never had a test for a sexually transmitted infection (STI):
56.2 per cent

who have sex at least once without a condom:
93.1 per cent

Proportion of surveyed British backpackers in Australia who had sex during their stay: **73.2 per cent**
who had multiple sex partners: **39.7 per cent**
proportion of these who were single: **45.7 per cent**

Proportion of surveyed Australians who refuse to mention contraception when having sex with a new partner:
63 per cent
of women: **52 per cent**

Proportion of surveyed Australian women who said they liked to drink alcohol before going to bed with their partner:
75 per cent
who have never had sex while sober:
6 per cent

Proportion of surveyed internet users in Australian who engaged in cybersex that were married or in a serious relationship:
55 per cent

Average age of male cybersex participants in Australia:
41 years
who had met someone online face-to-face after engaging in some sort of online sexual activity:

65 per cent

It's so long since I've had sex, I've forgotten who ties up whom.

Joan Rivers, US comedienne and actress

***

Proportion of surveyed 16-17-year-old boys in Australian who have watched pornography on the internet:
84 per cent
of girls: **60 per cent**

Proportion of surveyed British doctors who suggest having sex as a method of getting a good night's sleep: **17 per cent**
who suggest single beds: **38 per cent**

Proportion of surveyed US women who admit to having sex simply to receive gifts:
10 per cent
proportion of surveyed US female university students who sleep with a male friend who is not their boyfriend:
60 per cent

Level of noise recorded in a neighbour's flat that came from a British woman whenever she had sex:
47 decibels
year: **2009**

Proportion of Australian men who feel their religion dictates their choice of contraception:
16 per cent
of women: **6 per cent**

Proportion of surveyed single UK backpackers in Australia who had unprotected sex with multiple partners when here:
24 per cent

Proportion of Australian women who said they had been tipsy when sleeping with a partner for the first time:
40 per cent

Proportion of surveyed Australian butchers who report feeling healthy, laughing at work and think they are having more sex than other workers: **76 per cent** proportion of butchers who are actually having more sex than other workers: **60 per cent**

I blame my mother for my poor sex life. All she told me was, "the man goes on top and the woman underneath". For three years my husband and I slept on bunk beds.

Joan Rivers, US comedienne and actress

***

Average number of men that surveyed Australian women aged 18–50 had slept with: **8**
number of these they slept with when they were drunk: **5**
number of men whose names they couldn't remember the next day: **2**

Proportion of surveyed US women who admitted to having sex just to ensure a quiet life or to bargain for household chores:
84 per cent

Proportion of surveyed Australians who admit to secretly checking their partner's mobile phone messages:
1 in 3
proportion of these who found things they wish they hadn't:
73 per cent
who found messages ranging from the flirtatious to sexually explicit:

44 per cent

TECHNOLOGY

Why don't they make the whole plane out of that black box stuff?

Steven Wright, US comedian

Computers will never take the place of books. You can't stand on a floppy disk to reach a high shelf.

Sam Ewing, US writer and humourist

***

Proportion of surveyed mobile phone users in Australia who say they have sent text messages while driving:
25 per cent
who have taken their eyes off the road to look at their phone:
56 per cent

Proportion of surveyed Australians who regularly check their mobile phone statement:
53 per cent
age group least likely to check:
under 35 years

Estimated proportion of Australian households with Broadband at the beginning in 2008–9: **72 per cent**
with a computer: **78 per cent**

Number of active domain names worldwide in April 2010:
118 million

Estimated number of CCTV cameras in Britain in 2010:
about 4.2 million

by 2018: **8.6 million**

Estimated proportion of the world's population who have never placed or received a telephone call:
over 50 per cent

Factor by which the average desktop computer keyboard reportedly contains bacteria compared to a standard toilet seat:
400 times

Proportion of surveyed Australians who are very concerned about the unauthorised access or misuse of their personal information:
62 per cent
who are concerned about internet security and spam:
42 per cent

Estimated combined amount of time that Britons spend on hold waiting on customer telephone lines each year:
60 million hours

Estimated number of internet users in China at the beginning of 2010:
384 million

Estimated number of computers that end up in landfill each year in Australia:
1.6 million

Proportion of surveyed retirees in Australia who nominated their mobile phone as their preferred form of communication:
91 per cent

They've finally come up with the perfect office computer. If it makes a mistake, it blames another computer.

Milton Berle, US comedian and actor

***

Proportion of surveyed mobile phone users in Australia who say that their phone:
reduces their stress level: **31 per cent**
increases it: **10 per cent**
stays the same: **59 per cent**

Proportion of surveyed Australian mobile phone users who have been billed for more than $500 in a single month:
8 per cent

Proportion of Australian web visits that go to online shopping or classified sites:
5.7 per cent

Reported number of calls that are made to 000 each year:
about 12 million
proportion that come from mobile phones: **about 60 per cent**
proportion that are not an emergency: **60 per cent**

Reported average auctions each day on e-Bay Australia:
about 100,000
on Boxing Day 2008:
126,500

Estimated number of mobile phones that are reported lost or stolen in Australia each week:
3,200

Estimated proportion of software installed on computers worldwide in 2008 that was pirated: **41 per cent**
cost to the software companies worldwide: **about $US53 billion**
estimated proportion in Australia: **26 per cent**

Proportion of surveyed Australians who said they have used a mobile phone while on the toilet: **48 per cent**
Chinese: **66 per cent**

Proportion of surveyed Australians who think about recycling their old mobile phones:
25 per cent

Proportion of surveyed Australians who said they would use their mobile phone while eating a meal with others:
80 per cent

Reading computer manuals without the hardware is as frustrating as reading sex manuals without the software.

Arthur C. Clarke, British science fiction author and inventor (Clarke's 69th Law, The Odyssey File, 1984)

***

Reported average number of telemarketing calls received by Australians each month: **8.5**
estimated value of the time research, make calls each year: **$1.6 billion**

Proportion of surveyed online shoppers who do research, make comparisons or deliberate before making a purchase:
83 per cent

Estimated number of mobile phone services in Australia in 2008–9:
24.22 million
proportion of all internet subscriptions that are wireless:
25 per cent

Of a sample of 32 million user passwords on the web, the most common password: **123456**
second most: **12345**

third most: **123456789**

Chance that an unprotected personal computer will be infected by a virus within an hour of being connected to the Internet:
90 per cent

Proportion of surveyed Americans who feel that public mobile phone conversations are rude: **51 per cent**
not rude: **37 per cent**
not sure: **12 per cent**

Proportion of surveyed Australians who have sent a text message while at the movies: **34 per cent**
in church: **14 per cent**

Proportion of surveyed Australians who regularly swear at their mobile phone in frustration: **60 per cent**
who regularly misplace their phone and have to call to locate it:
77 per cent

Proportion of surveyed Australian men who secretly store phone numbers in their mobile to hide them from their partners:
25 per cent
women: **8 per cent**

Proportion of surveyed Australians who have accidentally sent a text message meant for their lover or partner to somebody else:
20 per cent

Estimated number of internet users worldwide in 2010:
about 1.80 billion

Year the first email was sent (by Roy Tomlinson, USA, with the message 'QWERTYUIOP'):
1971

The only way to make your PC go faster is to throw it out a window.

Robert Paul, British computer scientist and humourist

***

Estimated number of text messages sent by Australians on Christmas day 2009:
40 million
number of Christmas cards that were posted on or before that day:
70 million

Proportion of surveyed 18 to 31-year-olds in Australia who consider their mobile phone as their main communication tool:
70 per cent

Date the first Web site was launched:
6 August 1991

Proportion of Australians who know their best friend's phone number from memory: **33 per cent**
their own office number: **50 per cent**

Estimated number of spam email messages sent worldwide in 2009: **72 trillion**
proportion of all the world's email traffic that is spam:

92 per cent
estimated amount of power used sending, sorting and storing spam messages:
33 billion kilowatt hours

RELATIONSHIPS

I like long walks, especially when they are taken by people who annoy me.

Noel Coward, English playwright and composer

With my wife I don't get no respect. I made a toast on her birthday to 'the best woman a man ever had.' The waiter joined me.

Rodney Dangerfield, US comedian and actor

***

Proportion of surveyed married Australian women who feel it is essential to have a money-wise partner: **75 per cent**
of married men: **50 per cent**

Proportion of surveyed Australia women who feel they are treated differently by society because of their size:
45 per cent

Proportion of surveyed who say they have had an office romance at some time: **70 per cent**
proportion of these in which at least one of them was already in another relationship:
33 per cent

Proportion of people who kiss with their eyes open:
8 per cent

Proportion of surveyed British women who feel that a car journey is the perfect time to speak with their partner about big issues: **14 per cent**

proportion who feel that they are so busy they have to save important conversations for the car: **40 per cent**

Proportion of surveyed young Australian women who say they plan to live off their partner's income after they turn 50:
83 per cent
of young men:
73 per cent

Proportion of British men who fantasise about:
nurses: **44 per cent**
maids: **44 per cent**

air hostesses: **40 per cent**
traffic wardens: **6 per cent**

It was a woman who drove me to drink. I never had the courtesy to thank her.

W.C. Fields, US comedian and actor

***

Proportion of surveyed Australian workers who are in a workplace relationship at any one time: **11 per cent**
proportion of Australians who have had a relationship with someone at the office at some stage in their career:
80 per cent

Proportion of surveyed Australians aged 18–35 who said they had ended a relationship via SMS:
15.8 per cent

Reported number of facial muscles exercised during a passionate kiss: **34**
postural muscles: **112**

Proportion of surveyed Americans who say they have no close friends to confide in: **25 per cent**
in 1985: **10 per cent**

Proportion of surveyed single Australians aged over 18 who have used online dating services:
27 per cent

Proportion of surveyed Australians who believe that the way you look defines your character:
74 per cent

Proportion of surveyed engaged couples in Australian who said they fell in love at first sight:
65 per cent

Proportion of surveyed engaged couples in Australian who met:
through mutual acquaintances: **38 per cent**
while out clubbing: **13 per cent**
through family connections: **5 per cent**
online: **4 per cent**

Proportion of surveyed British women who say that shopping alone or with their children is nicer than with their partner:
89 per cent
proportion of these who cited their partner's bad behaviour as the reason:
76 per cent

Proportion of surveyed Australian men who say they feel comfortable when their wife or girlfriend is driving:
15 per cent

For every 100 single women aged 40 and over in Darwin looking for a partner in the same age bracket, the number of eligible men is: **105**
in Adelaide: **87**
in Sydney: **83**
in Melbourne: **81**
in Brisbane: **80**
in Canberra: **77**

Women: can't live with them, can't bury them in the back yard without the neighbours seeing.

Shaun Williamson, British actor, comedian and singer

***

Proportion of surveyed men worldwide who rated being in good health as the most important factor in quality of life:
33 per cent
having a harmonious family life: **26 per cent**
being in a good relationship with their partner: **19 per cent**
having a satisfying sex life: **2 per cent**

Proportion of surveyed Australians who said they would use a GPS to track their partners' whereabouts:
25 per cent
who would use their phone during extremely intimate moments: **13 per cent**

Average age at which a Chinese person reportedly has their first kiss:
23 years

When given the choice between sizes 8, 12 and 14, the proportion of surveyed Australian men who said they prefer a woman to be:

size 8: **20 per cent**
size 12: **41 per cent**
size 14: **39 per cent**

Average height of surveyed Australian women with an average age of 37: **162 centimetres**
average weight: **68 kilograms**
average bra size: **14C**
proportion who have been on a diet: **90 per cent**

Proportion of surveyed Americans who say they:
never lie: **11 per cent**
hardly ever lie: **68 per cent**
occasionally lie: **17 per cent**
often lie: **3 per cent**

Proportion of surveyed Americans who say the most fun way to flirt is in person: **55 per cent**
by text: **15 per cent**
by email: **3 per cent**

Estimated number of times an American woman shaves her legs in her lifetime: **6,626**
total hours spent doing so: **533**

Proportion of surveyed Australian women who said they had dated a co-worker:
19 per cent
who were still in a relationship with someone they met at work:

16 per cent
who had dated their boss:
6 per cent

I told my wife the truth. I told her I was seeing a psychiatrist. Then she told me the truth: that she was seeing a psychiatrist, two plumbers, and a bartender.

Rodney Dangerfield, US comedian and actor

***

Proportion of surveyed Australian women who said they have had a crush on a work colleague:
32 per cent

Maximum number of meaningful relationships the human brain can reportedly handle:
150

Proportion of surveyed Australians who rarely or never speak with their neighbours:
50 per cent
who say they are too busy to get to know their neighbours:
33 per cent
who know all their neighbours' names:
20 per cent

Estimated proportion of people who snore occasionally:
45 per cent
often: **25 per cent**

Proportion of surveyed American males who are happy with their weight: **67 per cent**
women: **54 per cent**

Proportion of surveyed British couples who said they have at least one argument each day over menial household tasks:
3 per cent
who spend at least one day each year refusing to speak with their partner after a housework row:
20 per cent

Estimated amount spent on chocolates and sweets on Valentine's Day in Australia in 2010:
$279.2 million
on weekend getaways:
$398.7 million

Number of single adults in Australia:
about 4.2 million
reported number of active users of online dating sites:
1.5 million

Most hugs given in a 1-hour period (Jeff Ondash, USA):
1,205
year: **2008**

Most hugs given in a 24-hour period (Jeff Ondash in Las Vegas on Valentine's Day 2010):

7,777

Proportion of surveyed Australian women who say they prefer a romantic dinner to having sex:
30 per cent

A man doesn't know what happiness is until he's married. By then it's too late.

Frank Sinatra, US actor and singer (The Joker is Wild, 1957)

***

Proportion by which the eye's pupil expands when a person looks at something pleasing:
up to 45 per cent

Proportion of British women who have hidden a new pair of shoes from their partner:
15 per cent
who have regretted at least one shoe purchase:
60 per cent

Proportion of surveyed British women who remember the name of the first boy they kissed: **66 per cent**
who felt bad about dumping a boyfriend: **15 per cent**

Proportion of surveyed New Zealand men for whom the first thing they look at in a woman is their:
breasts: **47 per cent**
waist and hips: **33 per cent**
face: **20 per cent**

Proportion of surveyed Victorians who would rather give up their partner than the internet:

70 per cent

Proportion of surveyed Australian fathers who say they always or almost always enjoy spending time with their children:
28 per cent
of mothers: **40 per cent**

Proportion of surveyed Australian couples who sleep in separate bedrooms because of a partner's noisy breathing or snoring:
20 per cent

Proportion of surveyed Australian women who ended a relationship over their partner's snoring:
10 per cent
who describe loud snoring as a turn-off:
70 per cent

Proportion of votes received by the book *"If You Want Closure in Your Relationship, Start With Your Legs"* in a competition for the oddest book title of 2007:
33 per cent
"I was Tortured By the Pygmy Love Queen":
20 per cent

Proportion of surveyed single Australian women in their twenties who say they are looking for a partner:
75 per cent

Proportion of surveyed Australians who feel that aliens are more likely to be friendly than not:
84 per cent

NOTABLE YEARS

I once spent a year in Philadelphia, I think it was on a Sunday.

W.C. Fields, US comedian

Remember, today is the tomorrow you worried about yesterday.

Dale Carnegie, US writer and lecturer

***

Year the last person was hanged in Australia (Ronald Ryan, Melbourne's Pentridge Prison): **1967**

Years that the Goon Show was first broadcast on the BBC radio:
1951–1960

Year the Hundred Years War ended: **1453**
time it lasted: **116 years**

Year that the TV sitcom *Father Knows Best* was first shown in the USA: **1954**
number of episodes made: **203**

Year of the Union of the Crowns (uniting the crowns of England and Scotland): **1603**

Years that Pythagoras lived: **about 569–500BC**
Aristotle: **384–322BC**

Year that Velcro was invented (by George de Mestral, Switzerland): **1948**

Year that the English model Jean Shrimpton wore her famous mini-skirt in Melbourne Cup week: **1965**
its height above her knees:
about 10 centimetres

Year that Wolfgang Amadeus Mozart died: **1791**
his age: **35 years**

Year that Christmas stamps were first issued by Australia Post (then the PMG): **1957**

Year that Meccano was invented and patented (by Frank Hornby, England): **1901**
Year the first person in the USA received a tax deduction, on appeal, for breast enlargements (Cynthia Hess, aka "Chesty Love", USA): **1988**
her new breast size: **56FF**
their individual weight: **4.5 kilograms**

amount claimed: **$US2,088**

The day after tomorrow is the third day of the rest of your life.

George Carlin (Sometimes a Little Brain Damage Can Help, 1984)

***

Year that the first person was killed by a motor car (Bridget Driscoll, London): **1890**
speed of the car: **6 kilometres per hour**

Year that New Zealand gave women the right to vote (the first major self-governing country in the world to do so): **1893**

Year the world's population reached:
1 billion: **1804**
2 billion: **1927**
3 billion: **1960**
4 billion: **1974**
5 billion: **1987**
6 billion: **1999**
estimated year to reach 7 billion: **2013**

Years of the Ming Dynasty in China: **1368–1644**

Year that the AIDS virus was first identified (in Atlanta, USA):
1981

Year that Braille was invented: **1829**
dynamite: **1866**
X-rays: **1895**
plastic: **1907**

Year that the Wright brothers flew the first powered aeroplane (named Kitty Hawk, in North Carolina, USA): **1903**
distance it travelled: **about 39 metres**

Year that John Quincy Adams became the first US president to have his photograph taken: **1843**

Year that Tungsten was discovered: **1783**
aluminium: **1825**
Krypton: **1898**

Years that Leonardo Da Vinci painted the Mona Lisa:
c.1503–1506
its dimensions: **77 centimetresx53 centimetres**

Year the first solar-powered hot water system was developed (by Roger Morse, Australia): **1953**

Year of last visit of the Comet Kohoutek: **1973**
next visit: **76,973**

Years that span the Jacobean period:
1603–1625

Year the Golden Gate Bridge in San Francisco was first opened:
1937

Year that Disneyland opened (in Anaheim): **1955**

Year that Murphy's Law was first coined (attributed to US Captain Ed Murphy, an engineer): **1949**

Yes, time flies. And where did it leave you? Old too soon ... smart too late.

Mike Tyson, US boxer

***

Year a book *(History of the War in the Peninsular and in the South of France from the Year 1807 to the Year 1814, Vol. I)* was stolen from Lexington's Washington and Lee University Library, USA: **1864**
year it was returned: **2009**

Year that Bugs Bunny first said, 'What's up, doc?' (in the cartoon *'A Wild Hare'*): **1940**

Year the first toilet cistern was seen on TV (in *Leave it to Beaver*):
1957

Years that Tug of War was an Olympic event:
1900 to 1920

Year Joan of Arc was burned to death: **1431**
her age: **19**

Year the Beach Boys were first formed: **1961**

Year of Shakespeare's birth: **1564**
his death: **1616**

Year the first electric battery was invented (by Alessandro Volta, Italy): **1800**

Year in which the Australian dollar was floated: **1983**

Year of the Battle of Agincourt: **1415**

Year the world's first modern police force was established
(in Paris): **1667**

Year of the Mutiny on the Bounty: **1789**

Year the Nobel Prize was first awarded: **1901**
number awarded that year: **5**
Years that Genghis Khan lived: **c.1162–1227**

Year the rack was introduced to the Tower of London (by the Duke of Exeter):
about 1420

Years that Guy Fawkes lived:
1570–1606

Years of the Aztec empire:
1427–1521

Year that Nero reportedly fiddled while Rome burned: **64AD**

Year that a belly button was first shown on television (by Yvette Mimieux in *Dr Kildare*): **1964**

Year that supermarket trolleys were first introduced (in Oklahoma, USA): **1937**

Age to women is like Kryptonite to Superman.

Kathy Lette, Australian author

***

Year that construction on Hadrian's Wall (in England) was finished: **122AD**
time it took to build: **about 6 years**
its length: **117 kilometres**

Year the Tang Dynasty was founded (by the Li family): **618**
year it ended: **907**

Year that King Canute came to the throne as King of England:
1016
King of Denmark: **1018**
year his reigns ended: **1035**

Year that William the Conqueror began building the Tower of London: **1066**

Year that work began on the Notre dame in Paris: **1163**
year it was effectively completed: **1345**

Year of the reign of Richard the Lionheart as King of England:

1189–1199

Year the first plane passenger was killed (Lieut. Thomas Selfridge in an aircraft piloted by Orville Wright): **1908**
Year Rowan Atkinson was born (in the UK): **1955**

Year the leaning tower of Pisa was closed for safety reasons:
1990
year it re-opened to the public: **2001**

Year the Dow Jones Industrial Average first closed over the 2000 mark: **1987**

Year that flirting in public toilets was outlawed in New York State: **1902**

Year that milk was first delivered in glass bottles (in New York): **1878**

Year of the first sale of Indian tea in Britain: **1839**

Year Agatha Christie was born (in Torquay, Devon, UK): **1890**
year she died: **1976**

Year the first VD clinic was opened in London: **1747**

Year that entertainer Liberace was born: **1919**

year he died: **1987**

Year that decimal currency was adopted in Australia: **1966**
in Britain: **1971**

Year Fidel Castro became prime minister of Cuba: **1959**

Year the first electronic tram ran down George Street, Sydney: **1899**
year the last tram ran through the streets of Sydney: **1962**

Tomorrow never comes. As soon as it does, it becomes today.

Anonymous

***

Year the foxtrot was invented (by Harry Fox, USA): **1914**

Year the religious reformer Martin Luther was born: **1483**
year he died: **1546**

Year that *Pilgrim's Progress* by John Bunyan was published: **1678**

Year King Louis XVI of France was executed for treason: **1793**

Year that Elvis Presley married Priscilla Beaulieu in Las Vegas: **1967**
year they divorced: **1973**
year he died: **1977**

Year of the Battle of the Alamo, a fort held by 180 Texans for 13 days against Mexican troops: **1836**

Year that New York mechanic Walter Hunt invented the safety pin: **1849**

Year that two leap days were added to Sweden’s Julian calendar to overcome dating discrepancies (and creating a February 30th): **1712**

MEDICINE

I told the doctor I broke my leg in two places. He told me to quit going to those places.

Henny Youngman, British-born comedian and violinist

Never go to a doctor whose office plants have died.

Erma Bombeck, US humourist

***

Proportion of cancers that are inherited:
5–10 per cent

Average weight gained by young Australian women each year:
649 grams
middle-aged women: **494 grams**

Proportion of surgeons in Australia who are aged over 55 years:
50 per cent
under 40 years: **16 per cent**

Proportion of blood flowing from the heart that is sent to nourish the brain:
20 per cent

Approximate sound level of a:
gunshot: **120 decibels**
noisy disco: **105 decibels**
chainsaw: **100 decibels**
ride-on mower: **90 decibels**
leaves rustling: **20 decibels**

Estimated number of asthma sufferers in Australia: **about 2.2 million**

Amount by which the risk of colon cancer is increased for men if they have a large waistline:
39 per cent
women: **48 per cent**

Estimated number of people in the world with diabetes in 2010:
about 285 million
in 2000: **171 million**
in 1985: **30 million**

Number of Australians with diabetes in 2010: **about 900,000**

Average number of people admitted annually to Dutch hospitals from injuries while playing darts:
about 120

Proportion of Australians affected with arthritis: **3.4 per cent**

Proportion of surgeons in Australia who are aged over 55 years: **50 per cent**
under 40 years: **16 per cent**

Average time in their life a male in Australia will spend with a severe disability: **5.4 years** a female: **8.3 years**

I just got out of the hospital. I was in a speed-reading accident. I hit a bookmark.

Steven Wright, US actor and writer

***

Average age at which breast cancer in women is diagnosed:
59 years

Average amount by which surveyed Australians underestimate their weight: **2.0 kilograms**
overestimate their height: **1.4 centimetres**

Estimated proportion of Australians with bad acne who spend more than $1,000 each year on treatment:
20 per cent

Proportion of the world's population with blood group O:
46 per cent

Estimated number of Australians affected by hearing loss:
3.55 million
proportion due to infection or injury: **17 per cent**
to age: **28 per cent**
to a birth defect: **4.5 per cent**

to prolonged loud noise/industrial deafness: **33.5 per cent**
to other causes: **17 per cent**

Number of major types of germs:
4 (bacteria, viruses, fungi, protozoa)

Proportion of surveyed Australian men who believe that breast cancer is more serious than prostate cancer:
23 per cent

Reported increase in the risk of developing cancer for males with a waistline over 100 centimetres or a female with a waistline over 85 centimetres:
13 per cent

Proportion of surveyed Australian men who said they only visited their doctor for a prostate check because of the prompting of a female friend or relative:
80 per cent

Proportion of surveyed 18–64 year-old Australian women who would consider plastic surgery to enhance their figure:
50 per cent

Average number of heart attacks suffered each hour in the UK:
30

average time sufferers wait before calling an ambulance:
90 minutes
proportion who die before reaching hospital:
1 in 3

Number of Australians each year who undergo skin grafts due to skin cancer:
about 45,000

When doctors and undertakers meet, they wink at each other.

W.C. Fields, US comedian and actor

***

Proportion of all blood donations in Australia that go to treating people with cancer:
30 per cent

Average length of a mid-sleep dream:
20–25 minutes

Proportion of total blood volume that is plasma:
about 55 per cent

Time it takes a mouthful of food to pass through the
oesophagus: **5–7 seconds**
stomach: **2–6 hours**
small intestine: **3–5 hours**
large intestine: **3–10 hours**

Number of dementia sufferers in Australia: **about 220,000**
proportion of these aged under 65: **5 per cent**
estimated number of sufferers by 2050: **about 730,000**

Proportion of Americans in 1965 who smoked:

42.4 per cent

Weight of a hairball found inside the stomach of an 18-year-old Chicago woman: **4.5 kilograms**

Volume of blood in a human body at any given time: **4.7–5.7 litres**

Proportion of surveyed new mothers in Australia who said they would consider cosmetic surgery to regain their figure:
70 per cent
who believe in aging gracefully:
13 per cent

Number of Australian adults diagnosed each year with a malignant brain tumour:
about 1,400
number each year who die from them:
about 1,200

Proportion of Australians who will experience Irritable Bowel Syndrome at some time:
about 20 per cent

Proportion of surveyed Australians who do not have regular dental check-ups:
50 per cent
whose last dental visit was at least 5 years ago:
12 per cent

who say that they cannot afford to go to the dentist:
35 per cent

Number of people admitted to US hospitals each year with knee injuries:
about 1.4 million

After two days in hospital I took a turn for the nurse.

W.C. Fields, US comedian and actor

***

Proportion of surveyed Americans aged 18–44 who take Aspirin daily: **4.6 per cent**
aged 45-64: **27.0 per cent**
aged 65 and over: **48.5 per cent**

Proportion of men in Australia who will develop prostate cancer:
1 in 9

Proportion of surveyed Australian women who prefer to have someone else in the room when being examined by a male clinician:
33 per cent

Proportion of surveyed Australians who say they trust their pharmacist:
90 per cent
who consider their pharmacist to be an expert in medicine:
82 per cent
who go to their pharmacist before or instead of their GP for some minor medical conditions:
66 per cent

Number of clinically diagnosable sleep disorders:
about 70

Average length of a yawn:
6 seconds
average interval between yawns in a bout of yawning:
68 seconds

High risk age group for Australian women to have breast cancer:
50–69 years

Proportion of 179 countries surveyed by the WHO that still allow smoking in hospitals and schools:
40 per cent
number that mandated pictorial warnings on tobacco packaging: 15

Estimated proportion of Australians who will be affected with a mind or brain disorder in their lifetime:
20 per cent

Number of organ donors in Australia in:
2009: **247**
2008: **257**
2007: **198**
2006: **202**
2005: **204**

Reported proportion of Australians who have had CPR training:
52 per cent
of people in Hong Kong: **12 per cent**

I tell you I don't get no respect. Why, the surgeon general, he offered me a cigarette.

Rodney Dangerfield, US comedian and actor

***

Number of people in Australia receiving organs from deceased donors in:
2009: **799**
2008: **846**
2007: **626**

Proportion of surveyed Australian adults who did not know that being overweight is a cancer risk:
51 per cent

Number of lap-banding surgeries performed in Australia each year: **over 8.000**
proportion done in private hospitals: **95 per cent**

Proportion of surveyed Australians who support organ donation:
94 per cent

Number of Australians who were waiting for a transplant in early 2010: **1,770**
number of these waiting for a kidney transplant:
1,310

Reported number of cancer patients in Australia who seek a second opinion about their disease and prognosis:
33 per cent

Reported number of Australians who are obese:
3.71 million
as a proportion of the population:
17.5 per cent
annual cost to the nation:
$58 billion

Maximum number per 10,000 people affected by a disease for it to be classified as rare: **5**
number of rare diseases:
between 5,000 and 8,000
proportion of Australians affected by them:
6–10 per cent

Proportion of surveyed Australian adults who have some sort of allergy: **44 per cent**
proportion of these who have experienced moderate or severe symptoms: **51 per cent**

Proportion of Australia's eligible donor population who give blood each year:
about 3 per cent

Length of the largest appendix removed (Safranco August, Croatia, during an autopsy):

26 centimetres
year: **2006**
his age: **72**
normal range in length: **7.5–10.0 centimetres**

Happiness is your dentist telling you it won't hurt and then having him catch his hand in the drill.

Johnny Carson, US talk show host

***

Proportion of blindness and vision impairment that can be prevented by early detection:
up to 80 per cent

Proportion of surveyed Australians who rated developing arthritis as their main medical concern:
43 per cent
proportion who will actually get it:
about 20 per cent

Estimated number of new and recurrent stroke victims in Australia in 2010: **60,000**
proportion of these that are to people aged under 55 years:
20 per cent

Number of PBS scripts written each year in Australia:
about 175 million
proportion of a visit to a General Practitioner that results in at least one script:
70 per cent

Proportion of surveyed Australians who say they do not have a long-term health problem: **23.3 per cent**
who have a skin disease: **3.8 per cent**
a urinary disease: **3.3 per cent**

Estimated number of Australians with Parkinson's disease in 2010: **about 100,000**
average age at diagnosis: **60 years**
proportion aged 40 or over: **90 per cent**

Proportion of new brain tumour patients in Australia who are aged less than 14 years:
20 per cent

Number of Australians each year who contract Chlamydia:
about 55,000
Americans: **about 3 million**

Factor by which the number of bacteria in your ear will multiply by wearing headphones for one hour:
700

Mortality rate from the pneumonic plague form of the Black Death in 1665: **90–95 per cent**
time it took symptoms to appear: **1–7 days**

Number of Australians each year who undergo skin grafts due to skin cancer:
about 45,000

Estimated number of smokers worldwide:
about 1 billion

Reported times more likely you will die of a heart attack if you are bored at work:
2.5 times

I broke my arm trying to fold a bed. It wasn't the kind that folds.

Steven Wright, US actor and writer

***

Proportion of men aged under 64 who have urinary incontinence: **5–7 per cent**
64 and over: **10–20 per cent**

Proportion of Australians aged over 45 who are affected by Chronic Pulmonary Disease (COPD): **up to 1 in 6**
proportion who do not know they have it: **75 per cent**

According to the World Health Organisation (WHO), the body mass index (BMI) that would classify you as:
normal: **18.5–24.9**
underweight: **16.5–18.4**
severely underweight: **less than 16.5**

Number of Australians with Parkinson's disease: **about 55,000**
proportion who are diagnosed before age 45: **about 10 per cent**

Estimated proportion of Australians aged over 65 who have moderate to severe dementia: **1 in 15**
aged 80 to 84: **1 in 9**
85 and over: **1 in 4**

Number of reported cases of dengue fever in Singapore in 2009: **7,031**
in 2007: **8,826**

Proportion of surveyed Australian women who want a female doctor only for an anogenital examination:
64 per cent

Proportion of surveyed Australians who have given up taking a medication before the end of the recommended course:
41 per cent
proportion of these who did so because of perceived unwanted side effects: **42 per cent**

Number of deaths worldwide from measles in:
2008: **197,000**
2004: **454,000**
1999: **871,000**

Proportion of blind people who are born without sight:
about 35 per cent

Estimated proportion of nursing home residents in Australia who are affected by incontinence:
about 77 per cent

Number of male cancers in Australia that are breast cancer:
less than 1 per cent

Number of GPs per 100,000 people in Australia in:
2006: **178.6**
2000: **191.5**
1986: **152.6**

I just got the bill for my operation. Now I know why those guys wear masks.

Jim Boren, US humourist and author

***

Reported number of objects (such a pliers, scalpels, coils and swabs) mistakenly left inside patients by Britain's National Health Service in 2008:
722

Energy used to bang your head against a wall for one hour:
628 kilojoules

Proportion of surveyed American women who say they have at least one headache a year: **95 per cent**
of men: **90 per cent**

Estimated proportion of people who have sucked their thumbs:
80 per cent

Estimated number of bacteria living on Earth:
5 million trillion trillion

Number of human skulls in the vaults of the Chapel of Skulls in Czermna, Poland: **over 20,000**
year it was created: **1776**

Proportion of surveyed Australians who had delayed or avoided visits to a GP, dentist or specialist doctor in the previous six months due to financial problems:
28 per cent

Number of Americans admitted to hospitals in 2007:
39.5 million
in 1997: **34.7 million**

Proportion of Hong Kong's population in 2007 who smoked:
11.8 per cent
in 2005: **14 per cent**

Estimated proportion of middle aged Australians who are at a healthy weight: **25 per cent**
annual cost of caring for the nation's overweight and obese people: **$56 billion**

Proportion of surveyed Australians who find nurses ethical and honest (highest ranking): **89 per cent**

pharmacists: **84 per cent**
doctors: **82 per cent**
business executives: **16 per cent**
car salesmen (lowest ranking): **3 per cent**

I saw a specialist who asked me: Are you familiar with the phrase 'faecal impaction'?' I said: 'I think I saw that one with Glenn Close and Michael Douglas'.

Bob Monkhouse, British entertainer

***

Reported proportion of people worldwide each year that suffer from typhoid fever: **16–33 million**
age with the highest incidence: **5–19 years**

Number of Australian women diagnosed with ovarian cancer in 2010: **about 1,500**
proportion who are diagnosed in the advanced stages and will not live beyond 5 years: **75 per cent**

Number of Australians who die of skin cancer each year:
about 1,700
proportion of surveyed teenagers who think a tan looks
healthy:
43 per cent
who admit to getting burnt on an average summer weekend:
25 per cent

Number of Australians who receive some sort of aged care:

about 1 million
number of nursing homes in Australia:
about 2,800
proportion of residential places occupied in 2008–9:
92.9 per cent

THE HUMAN BODY

I'm so old they've cancelled my blood type.

Bob Hope, US comedian

I tell you nothing goes right. My wife's father died. She had his body frozen. Every time I take a snack, he falls out of the refrigerator.

Rodney Dangerfield, US comedian and actor

***

Reported proportion of women in the USA who are obese:
34 per cent
France: **11 per cent**
Japan: **3 per cent**

Area of an adult human's skin: **1.3–2.0 square metres**
minimum thickness: **0.5 millimetres (eyelid)**
maximum thickness: **4 millimetres (palm of hand)**

Amount of blood a human heart pumps in a day:
6,000–7,500 litres

Proportion of muscle men lose between the ages of 30 and 80 years:
40 per cent

Diameter of a human eyeball:
at birth: **17–18 millimetres**
at puberty: **20–21 millimetres**
in adulthood: **24 millimetres**

Number of neurons in a human brain:
100 billion

Average length of the world's longest fingernails on both hands (Lee Redmond, UK):
85 centimetres
combined length of all her nails: **8.54 metres**
year they were broken off (in a car accident): **2009**

Number of different odours a human can distinguish between:
about 10,000

Length of time that a human blush can last:
up to 5 minutes

Average number of hairs on the head of a naturally blond
woman: **about 120,000**
brunette: **about 100,000**
redhead: **about 80,000**

Typical lifespan of a human hair:
4–8 years

I tell you, I'm not a sexy guy. I was with one girl, I said to her, "Come on honey, I'll show you where it's at." She said, "You'd better, 'cause the last time I couldn't find it."

Rodney Dangerfield, US comedian and actor

***

Average amount of water lost through the skin each day by an adult male at rest:
15 millilitres

Time it takers an unembalmed body buried two metres down in ordinary soil to decompose down to the skeleton:
10–12 years
for the bones to fully decompose:
about 300 years

Average weight of a human eyeball:
28 grams

Estimated number of human protein-coding genes:
20,000–25,000

Amount of skin a human sheds in an average lifetime:
about 18 kilograms

Number of bacteria in a human gut: **about 750 trillion**

Number of red cells in a drop of male blood: **4.5–6.2 million**
female: **4.2–5.4 million**

Number of primary taste sensations:
4 (salt, sweet, sour, bitter)

Average time an adult can hold their breath:
about 1 minute

Proportion of people who lose their sense of smell who at the same time lose their sex drive:
25 per cent

Average amount of saliva produced by the mouth each day:
1–2 litres
in a lifetime:
between 27,000 and 40,000 litres

Number of bacteria per square millimetre under a human armpit:
about 203,000

pH range of the acidity of a human stomach:
1.5–3.5

Number of muscles used in the face and neck to say one word:
over 70

Average number of cells per square centimetre of human skin:
7.4 million

Thickness of a strand of human hair:
about 0.008 millimetres

Happiness is good health and a bad memory.

Ingrid Bergman, Swedish actress

***

Number of glands that secrete wax in a human ear:
about 2000

Proportion of human taste buds that are lost by age 60:
50 per cent
life span of a taste bud:
7–10 days

Average number of individual bacteria living on one square centimetre of human skin:
10 million
number of sweat glands: **256**
hairs: **24**
sensory cells: **7,480**

Lifespan of a skin cell:
about 35 days

Time it takes blood vessels damaged by sunburn to return to their normal state:
up to 15 months

Normal body temperature of a human:

about 37 degrees centigrade

Proportion of human hair that is protein:
91 per cent

Average number of hairs a human loses each day:
about 200

World's longest hair (Xie Qiuping, China):
5.627 metres
year measured: **2004**

Number of bones in a human big toe: **2**
in each of the other toes: **3**

Number of times that an average human heart beats in one minute: **72**
in a day: **about 104,000**

Average weight of human skull:
about 3.2 kilograms

Time it takes a red blood cell to circle the whole human body:
about 20 seconds

Number of digestive glands in a human stomach:
35 million
time it takes food to get from your mouth to your stomach:

about 7 seconds

Common looking people are the best in the world: that is the reason the Lord makes so many of them.

Abraham Lincoln, former US President

***

Number of muscles used by a human when balancing:
about 300

Average height of an adult Australian woman:
163.9 centimetres
man: **178.4 centimetres**

Amount of water a body sweats per hour in a hot climate:
up to 500 millilitres

Length to which Kim Goodman, USA, could protrude her eyeballs beyond her eye sockets:
12 millimetres
year: **2007**

Reported world's longest pubic hair measured from the groin (Maoni Vi, South Africa):
71 centimetres
length of the hair from her armpits:
81 centimetres

Proportion of house dust that is due to skin cells shed by humans:
up to 70 per cent

Proportion of 75-year-olds with 20–20 vision:
about 15 per cent

World's longest moustache (Ram Singh Chauhan, India):
3.50 metres
year: **2008**

Proportion of surveyed Australians who admit to rarely eating healthy food and consuming too much junk food:
17 per cent

Estimated number of chemicals the average British woman hosts on her body every day (mostly from deodorants,
moisturisers and lipsticks): **515**
proportion who said that this does not worry them:
70 per cent

Longest female legs (Svetlana Pankratova, Russia):
132 centimetres
year: **2008**

Estimated proportion of Australians who smoke tobacco or cigarettes on a daily basis:

19 per cent

Number of bones in a human wrist: **8**

Number of nerve endings per centimetre on a human hand:
about 200

THE ANIMAL KINGDOM

HAM AND EGGS – A day's work for a chicken but a lifetime commitment for a pig.

Anonymous

All animals are equal but some animals are more equal than others.

George Orwell, English author and journalist

***

Reported number of shark attacks worldwide in:
2009: **61**
2008: **60**
2007: **71**
2004: **61**
number that were fatal in 2009: **5**

Number of species of kangaroo:
over 60
number of kangaroos in Australia:
30–50 million

Number of species of birds in Australia:
about 800
proportion that are unique to Australia:
50 per cent

Amount of water a mature elephant can carry in its trunk:
up to 4–6 litres
amount it drinks each day:
80–160 litres

Number of tentacles on the box jellyfish: **up to 60**
length of each tentacle: **up to 3 metres**
number of stinging cells per square centimetre:
up to 1 million

Number of egg-laying mammals:
2 (platypus and echidna)

Amount paid by an Alaskan zoo in 2005 to build a treadmill for a 3.6 tonne elephant:
about **$A200,000**
number of times the elephant has used it so far:
0

Proportion of surveyed British veterinarians who say they have treated an intoxicated dog:
25 percent

My father told me all about the birds and the bees. The liar – I went steady with a woodpecker till I was twenty-one.

Bob Hope, US actor and comedian

***

Longest recorded tip-to-tip horn span of a domestic goat (named "Uncle Sam", USA):
132 centimetres
year: **2004**

Distance from which a silkworm moth can reportedly detect another moth:
up to 11 kilometres

The distance an Arctic tern travels when migrating from one pole to another and back again (world's longest migration):
about 35,000 kilometres

Time that a rat can tread water:
up to 3 days

Time that a loggerhead turtle can hold its breath underwater:
up to 10 hours

Estimated number of house cats worldwide:

about 600 million

Number of arms on a cuttlefish: **8**
tentacles: **2**

Age at which *Tyrannosaurus Rex* reached full skeletal maturity:
about 20 years
proportion that lived to full maturity: **2 per cent**
its maximum length: **up to 13 metres**
its maximum weight: **up to 6.8 tonnes**

Number of dairy cows in New Zealand in 2009:
about 5.6 million
size of the average herd: **351**
average annual production of milk per cow:
3,567 litres

Amount of extra milk each year cows with names such as Daisy, Gertrude or Buttercup can reportedly produce over cows with no name:
284 litres

Amount paid by an American woman to a South Korean company to clone her dead pit-bull terrier 'Booger':
$US50,000
year: **2008**

Number of eyelids a camel has on each eye: **3**

Proportion of male giant pandas that are sterile:
90 per cent
of females that cannot bear cubs:
78 per cent

Number of teeth most toads have: **0**

I don't get no respect at all. My wife, she ran off with my best friend. Now I got no dog.

Rodney Dangerfield, US comedian and actor

***

Average life expectancy of a mockingbird:
10 years

Estimated proportion of Australians who are bitten by an animal at least once in their lifetime: **50 per cent**
proportion of these that are dog bites:
about 90 per cent

Estimated number of people poisoned by snakebites worldwide each year: **over 400,000**
number of these who die: **about 20,000**

Amount reportedly inherited by two dogs in Melbourne (a cairn terrier cross and a corgi) in a will by their owner:
$700,000
year: **2008**

Top speed of the Australian dragonfly (the fastest flying insect):
58 kilometres per hour

Proportion of time a three-toed sloth spends sleeping or dozing:
up to 80 per cent

Distance a typical rat can swim without resting:
up to 600 metres

Proportion of surveyed cat owners in Briton who believe their cat can open doors:
33 per cent
say "hello": **29 per cent**
who sit their cat at the dinner table to be fed:
16 per cent

Average heartbeat per minute of a humming bird:
1,300
a human: **72**
a blue whale: **10**

Length to which a saltwater crocodile can grow:

up to 7 metres

Amount of water a camel can drink in three minutes: **up to 200 litres**

Oh, I was an ugly kid. My old man took me to the zoo. They thanked him for returning me.

Rodney Dangerfield, US comedian and actor

***

Number of humans that could be killed by the poison contained in one Australian sea wasp (box jellyfish): **60**
time to death after being stung: **from 4 minutes**
their length: **up to 3 metres**

Number of species of rodents: **about 1,700**
bats: **about 1,000**
monkeys: **about 250**

Average life expectancy of a worker bee (in the summer months): **42 days**
drone: **90 days**
queen bee: **3–5 years**

Height of an ostrich: **up to 2.7 metres**
weight: **up to 160 kilograms**
top speed: **up 70 kilometres per hour**

Proportion by which the world's tiger population has declined in the past 100 years: **95 per cent**
estimated number now left in the wild: **5,000–7,500**

Number of rows of whiskers a cat has on each side of its face:
4

Number of known Pinta Island Tortoises:
1 (named "Lonesome George")
his age: **60–90 years**

Top recorded speed of the cosmopolitan sailfish (world's fastest over short distances):
109 kilometres per hour

Proportion of polar bears that live in Canada:
60 per cent
normal life span: **up to 18 years**
oldest life span in captivity: **41 years**

Estimated number of wild rabbits in Australia:
about 300 million
year they were introduced into the wild: **1859**
number released at that time: **24**

Number of sheep and lambs in Australia in 2009–10 (the lowest since 1905):
72.7 million

Average lifespan of a house fly:
15 to 30 days

Time it takes an ostrich egg to hatch:

about 42 days
an emu egg: **about 56 days**

Our perfect companions never have fewer than four feet.

Colette, French novelist

***

Proportion of the weight of an African elephant that is due to its skeleton: **about 15 per cent**

The proportion of a hummingbird's weight that is due to its brain (the largest proportion of any bird):
4.2 per cent

Number of brains a leech has: **32**
number of known species: **650**

Number of dogs reportedly on board the Titanic: **10**
number known to have survived: **3**

Most recorded calves born at a single birth (in Reysona, Tamaulipas, Mexico): **5**
year: **2005**

Highest recorded jump by a pot-bellied pig (named "Kotetsu" in Mie, Japan):
70 centimetres
year: **2004**

Leg span of a huntsman spider: **up to 15 centimetres**
lifespan: **about two years**

Longest ears on a dog ('Tigger', a bloodhound, USA):
left ear length: **34.2 centimetres**
right ear: **34.9 centimetres**
year: **2004**

Proportion of surveyed dog owners who regularly kiss them:
63 per cent

Number of species of horned lizard found in North America:
13
distance they can shoot blood from one or both eyes:
up to 2 metres

Estimated number of people worldwide who have one or more Ascaris worms living inside them:
1.4 billion

Length of a male adult roundworm: **15–30 centimetres**
female: **20–35 centimetres**

Range of a skunk's spray: **up to 4.5 metres**

Number of different types of cockroach:
more than 4,000
length of the wingless Australian rhinoceros cockroach:
up to 8 centimetres
its weight: **up to 33.5 grams**

Average length of a hippopotamus (including a tail of about 56 centimetres): **3.3 to 5.2 metres**
average weight of a:
male: **1,500–1,800 kilograms**
female: **1,300–1,500 kilograms**

Eagles may soar, but weasels don't get sucked into jet engines.

John Benfield, British actor

***

Typical length of a sea cucumber:
1–30 centimetres

Length of an Amazon giant centipede:
about 30 centimetres
number of legs: **46**

Reported distance a silver carp can leap from the water:
up to 3 metres

Number of species of wolf spider: **more than 2,000**
size (including their legs): **between 1 and 30 millimetres**
number of eyes: **8**

Length of the brown recluse spider: **5–20 millimetres**
number of eyes: **6**

Number of species of cicada: **about 2,500**
typical length: **2–5 centimetres**

Length of the tarantula hawk wasp: **up to 5 centimetres**
wingspan: **up to 7.5 centimetres**

Length of a death watch beetle:
about 7 centimetres

Body length of the Japanese spider crab (the world's largest):
up to 40 centimetres
length of its outstretched leg: **up to 3.8 metres**
weight: **up to 19 kilograms**
typical depth at which it can withstand water pressure:
up to 600 metres

Length of the penis of a sperm whale: **1.70 metres**
a hamster: **2 millimetres**

Proportion of surveyed American cat owners who say their pet sleeps in a family member's bed each night: **51 per cent**
dog owners: **36 per cent**

Gestation period of a koala: **about 35 days**
length of a joey when born: **about 60 millimetres**
weight: **less than 1 gram**

Proportion of human DNA shared by
chimpanzees: **about 98 per cent**
dogs: **75 per cent**

fruit lies: **60 per cent**
bananas: **50 per cent**

Number of birds in Jurong Bird Park in Singapore (the world's largest): **over 8,000**
number of different species: **about 600**

Number of species of goat: **8**

Women are like elephants to me. I like to look at them, but I wouldn't want to own one.

W.C. Fields, US comedian and actor

***

Number of species of ferns: **10,000**
algae: **12,000**
fungi: **100,000**

Natural life expectancy of a lobster:
50–100 years

Estimated proportion of creatures on Earth that have 6 legs:
80 per cent

Life expectancy of a flamingo:
about 40 years

Estimated number of feral pigs in Australia:
about 23 million

Number of chickens killed in Australia during 2009:
about 469 million

Number of legs on a butterfly: **6**

Length of time ‘Mike’, a chicken, survived without a head:
18 months
years: **1945–6**

Longest tail on a horse (mare JJS Summer Breeze):
381 centimetres
year measured: **2007**

Number of humans that could be killed by the toxin contained in a single bite of the inland taipan:
about 100

Estimated number of pet dogs in the USA in 2009:
78 million
cats: **94 million**
birds: **15 million**
freshwater fish: **172 million**

Life expectancy of a King cobra:
up to 20 years

Weight of the brain of an elephant:
about 6 kilograms
a cat: **about 30 grams**

Distance from which a falcon can see a 10 centimetre object:
1.5 kilometres

Reported number of words known by a budgerigar (named "Puck", USA): **1,728**
his age at death: **5 years**

Weight of a badger: **4.6–7.3 kilograms**
total length: **65–80 centimetres**

MONEY AND FINANCE

How to make a million dollars: first, get a million dollars.

Steve Martin, US actor and comedian

A bank is a place where they lend you an umbrella in fair weather and ask for it back when it begins to rain.

Robert Frost

***

Proportion of Australian adults who receive an inheritance in any one year:
1.4 per cent

Number of German marks to UK£1 in:
July 1914: **20**
January 1921: **315**
January 1922: **932**
July 1923: **1.7 million**
August 1923: **22.5 million**
September 1923: **480.5 million**
November 1923: **20 billion**

Estimated annual cost to Australian business due to illness:
about $7 billion

Selling price of the first Holden car produced in Australia: **£760**
year: **1948**

Proportion of surveyed Australians who say they have an incompetent or sexist boss:
82 per cent
proportion of these who complained directly to the boss about it:
8 per cent

Worldwide revenue from online gaming in 2010:
about $US18 billion

Number of bank notes produced each day in the USA:
about 35 million
their face value: **$US635 million**
proportion that are $1 notes: **45 per cent**

Proportion of Australian beef production that is exported: over 60 per cent
the value of exports: **$A5.4 billion**

Amount of wine exported from Australia each year:
670 million litres
its value: **$A2.7 billion**

Year the Commonwealth Bank of Australia began as a savings bank: **1912**
commercial bank: **1913**

Proportion of bankrupts in Australia who are aged under 30 years: **20 per cent**
30–39 years: **28 per cent**

40 years or older: **52 per cent**

Estimated number of Australians with multiple superannuation accounts: **4.3 million**
total number of accounts: **28 million**

Women will never be as successful as men because they have no wives to advise them.

Dick Van Dyke, US comedian, actor, writer and producer

***

Number of Australians each year who suffer a work-related injury or illness:
about 700,000

Amount paid at a London auction in July 2006 for the first guitar Beatle Paul McCartney reportedly ever played:
£330,000

Record price paid in June 2007 for a licence to operate a taxi cab in New York:
about **$US600,000**
cost in 2001: **$US195,000**

Weight of the world's largest coin (by the Royal Canadian Mint):
100 kilograms
thickness: **1 inch**
face value: about **$A1.1 million**
year: **2007**

Estimated amount Americans spend each year on weight-loss products and services:
over $US60 billion

Cost of a first-class return trip from Melbourne to England on then P & O liner *Mooltan* in 1932: **$344**
tourist class: **$140**

Proportion of surveyed Americans aged 50 and over with an annual income less than $A65,000 who say they believe in Heaven: **90 per cent**
more than **$A65,000: 78 per cent**

Amount paid at auction for a white truffle (found in Italy) by Stanley Ho, Macau:
about $A290,000
its weight: **about 1.5 kilograms**
year: **2007**

amount of money laundered each year in Australia:
about $4.5 billion

Amount paid by an 89-year-old English widow for a taxi ride to and around Greece from the UK: **about $A4,750**
distance travelled: **about 6,500 kilometres**

GREECE
UK
TAXI

Giving money and power to government is like giving whiskey and car keys to teenage boys.

P.J. O'Rourke, US political satirist, journalist, writer and author

***

Proportion of surveyed Australian workers who say their boss is completely ineffective:
58 per cent

Proportion of surveyed Australian men who feel that poor people are trapped in poverty: **39 per cent**
Americans: **29 per cent**

Number of Britons aged 15–24 who are not in work, education or training:
1.24 million

Proportion of surveyed Australian business managers who say they do not like, respect or enjoy working with their staff:
20 per cent

Record amount paid in 2006 for a Stradivarius (the "Hammer" at an auction in New York): **$US3.5 million**
year it was made: **1707**

Amount paid at auction in April 2008 for an 1864 antislavery letter written by Abraham Lincoln:
$US3.4 million

Amount spent on male cosmetics in Australia in 2007:
about $44.2 million

Estimated number of Americans living in the USA in 2008 who were in poverty: **39.1 million**
the poverty threshold annual income for a family of four:
$US22,050

Amount paid at auction in August 2008 for a piece of wedding cake from the wedding of Prince Charles to Lady Diana Spencer in 1981: **£1,000**
its weight: **800 grams**

Amount paid at auction in Japan for a jumbo black skinned watermelon in June 2008: **$US6,100**
its weight: **7.7 kilograms**

Average size of the loan for a first home buyer in Australia in 2009:
$281,700

Amount paid for a handwritten copy of the revised Magna Carta at a December 2007 New York auction:

$US21.321 million
year it was issued by King Edward I: **1297**

Proportion of surveyed Americans who wish their paper money was different colours for easier identification:
68 per cent

Proportion of surveyed senior executives in the USA who said they would hire a qualified 72-year-old as a CEO:
53 per cent

Government's view of the economy could be summed up in a few short phrases: If it moves, tax it. If it keeps moving, regulate it. And if it stops moving, subsidise it.

Ronald Reagan, actor and former US President

***

Amount paid at a French auction in June 2008 for a beige soapstone seal of the Chinese emperor Kangxi (c. 1661–1772):
$US7.2 million
its weight: **3 kilograms**

Winning bid at the June 2008 auction in Sydney for Picasso's painting *Sylvette* (an Australian record):
$5.75 million
year it was painted: **1954**

Number of books purchased in Australia in 2007:
about 63 million
amount spent on them: **about $1.25 billion**

Amount paid at auction in March 2009 for an envelope believed to be the earliest known airmail item addressed to Australia:
$190,800
year it was sent: **1870**

Proportion of Australian business owners who are male:
68 per cent

Amount Australians spent on toilet paper in 2007:
$728 million
total weight: **120,000 tonnes**
proportion sold that is premium grade (thicker and softer):
about 67 per cent

Reported inflation rate in Zimbabwe in October 2008:
231 million per cent
date the $100 trillion note was introduced:
16 January 2009

Amount paid at a London auction in February 2006 for a bottle of vintage Chateau d'Yquem: **$US90,000**
year it was bottled: **1787**

Amount paid at a UK auction in October 2006 for a lifejacket from the Titanic:
£43,000

Amount paid to American Indians for Manhattan Island in 1626:
$US24

Average amount Australians spend each year on their credit cards:

$14,310

Estimated average value of unwanted clothing in the wardrobes of Australian women: **$725**
men: **$327**

If women didn't exist, all the money in the world would have no meaning.

Aristotle Onassis, Greek shipping magnate

Amount fetched at auction for a painting by Norman Rockwell 'Extra Good Boys and Girls': **$US2.17 million**
when: **December 2007**

Number of ATMs in Australia in 2010: **about 24,600**
in 1997: **about 8,800**

Estimated proportion of Australians who received an unwanted Christmas gift in 2007: **65 per cent**
amount spent on them: **about $985 million**

Proportion of surveyed 18–25 year-old Americans who feel that getting rich is the first or second most important life goal:
80 per cent
being famous: **51 per cent**

Proportion of NSW householders in the highest income bracket:
37.2 per cent
Victoria: **23.4 per cent**
Queensland: **17.6 per cent**

Amount paid at auction in Japan in May 2008 for a pair of cantaloupes from Yubari:
$US23,500

Amount paid at a July 2008 auction for a baseball cap worn by Babe Ruth:
$US328,000

Estimated amount that wealthy Americans are reportedly hiding in overseas tax havens:
about $US1.5 trillion
global figures: **up to $US7 trillion**

Reported list price of a Qantas A380 aircraft in 2008:
$380 million – $386 million
maximum take-off weight: **569 tonnes**
number of passengers it can hold (in four classes): **450**

Amount paid at a May 2008 auction in Hong Kong for a 101-carat colourless diamond (the size of a squash ball):
$US46.2 million

Estimated cost each year to Australians of identity fraud:
$3.5 billion

Estimated value of unused annual leave on the books of Australian businesses in 2010:
$33.3 billion
number of days: **about 123 million**

Proportion of surveyed Australians who say they rarely tip:
56 per cent

Here's something to think about: How come you never see a headline like 'Psychic Wins Lottery'?

Jay Leno, US talk show host

***

Average weekly ordinary time earnings by full-time adult employees in Australia in February 2010: **$1,244**
for males: **$1,331**
females: **$1,095**

Estimated cost to the Australian economy each year of vision disorders: **$9.85 billion**

Median monthly home repayments in Australia in March 2009:
$1,831
in 1976: **$459**

Estimated amount spent each year in Australia spent on Christmas gifts: **$8.36 billion**
total number of unwanted gifts: **19 million**
their value: **about $978 million**

Amount paid (including a $900,000 buyer's premium) in March 2010 at an Australian auction for a Sydney Nolan painting of Ned Kelly:
$5.4 million

Proportion of surveyed Americans who see China as an economic threat:
72 per cent

Proportion of surveyed online shoppers who will make an impulse buy if they think they are getting a bargain:
76 per cent

Estimated weight of receipt paper that is generated by American retailers each year:
104 million kilograms

Reported annual salary in 2009 of the coach of the University of Southern California football team:
$US4.4 million
average salary of top level US football coaches:
$US1.36 million
increase since 2006: **46 per cent**

Reported losses of the US Postal Service in the financial year ending 30 September 2009: **$US3.8 billion**
total mail volume: **177.1 billion pieces**

Amount paid at a 2010 online auction for the first comic to feature Superman (Action Comics No. 1):
$US1.5 million
original cost of the comic: **US10 cents**

Reported wealth of the world's richest person (Carlos Slim Helu, Mexico) in March 2010:
$US53.5 billion
Bill Gates: **$US53 billion**
Warren Buffett: **$US47 billion**

Whoever said money can't buy happiness didn't know where to shop.

Gertrude Stein, US writer

***

Proportion of surveyed Australian women who do not trust their partner with their shared finances:
19 per cent
who believe that their partner wastes money:
47 per cent

Amount fetched at auction in 2010 for a 1939 comic book in which Batman made his debut (Detective Comics No 27):
$US1,075,500
amount it sold for in the late 1960s: **$US100**

Proportion of surveyed Australian women who say they do not save money on a regular basis:
56 per cent

Amount fetched at a 2010 UK auction for an unfinished cigar smoked by Winston Churchill: **£4,500**
for a butterdish he used as an ashtray: **£4,200**

Proportion of Australians aged 45–59 who believe that they will be fully reliant on the age pension: **12 per cent** partly reliant: **46 per cent**

Income generated by overseas students in Australia in 2008–9:
$16.6 billion

Proportion of surveyed Australians who say they would spend a tax return on reducing their debt: **53.4 per cent**
on body enhancements: **3.3 per cent**

Proportion of the world's 100 richest men in early 2008 who whose wife's hair colour was:
brunette: **62 per cent**
blonde: **22 per cent**
black: **16 per cent**
red/copper: **0 per cent**

British record price paid at auction for a Victoria Cross (awarded in 1943 to 21-year-old Briton Bill Reid): **£335,000**
date of the auction: **November 2009**
the world record price for a VC (in Sydney in 2006): **£491,567**

Average mortgage size in Australia in April 2010: **$303,200**
average credit card debt: **$2,108**

Amount paid at a December 2009 auction for a Rembrandt painting 'Portrait of a man, half-length, with his arms akimbo':
£20.2 million
time since it had been seen in public: **40 years**
year it was painted: **1658**

Proportion of surveyed Australian men who feel that income is largely determined by luck:
40 per cent
Americans: **30 per cent**

It isn't necessary to be rich and famous to be happy. It's only necessary to be rich.

Alan Alda, US actor, director and screenwriter

***

Year the median price of a house in Sydney reached:
$100,000: **1986**
$200,000: **1994**
$300,000: **1999**
$400,000: **2001**
$500,000: **2003**
$600,000: **2009**

Estimated number of gifts the average Australian child receives at Christmas: **12**
their value: **$277**

Proportion of surveyed Australian women who say they charge five or more purchases each week on credit cards:
19 per cent

Amount paid at auction in October 2009 for a flag believed to be the only surviving Union Jack from the Battle of Trafalgar:
£384,000
year of the battle: **1805**

Estimated proportion of individual tax payers in Australia who use a tax agent:
75 per cent

Proportion of US banknotes that show traces of cocaine in 2009:
90 per cent
in 1989: **67 per cent**

Amount paid for a US 1794 Liberty dollar in May 2010 (making it the world's most expensive coin):
$US7.85 million

Proportion of US coins that were 1 cent pieces (pennies) in 2009:
66 per cent
in 2007: **51 per cent**

Estimated value of the worldwide online gaming market by 2012: **$32 billion**

Proportion of surveyed people in London who would stoop to pick up a dropped penny: **25 per cent**
in south east and south west England: **44 per cent**
Scotland: **43 per cent**

Number of Britons who live in a household where nobody goes to work:
about 6 million

Amount paid for a cave home named ‘Rock Cottage’ at auction in west central England in July 2007: **£100,000**
time it was last inhabited: **1948**
when it was built: **1770s**

THE WORKPLACE

A good rule of thumb is if you have made it to thirty-five and your job still requires you to wear a name tag, you've made a serious vocational error.

Dennis Miller, US comedian

Employee of the month is a good example of how somebody can be both a winner and a loser at the same time.

Demetri Martin, US comedian, actor, artist, musician, writer and humorist

***

Proportion of those who experience conflict between work and home life who want to work more hours:
67 per cent

Average length of a worker's smoke break: **5 minutes**
coffee break: **10 minutes**

Proportion of Australian workers who are over-qualified for their job:
27 per cent

Average number of sick days taken by Australians each year:
10.4
estimated proportion of these days that were non-genuine:
10–25 per cent

Proportion of Australian workers who say their bosses delegate work outside their official job description:
67 per cent

Proportion of time the average Australian office worker spends sitting down: **77 per cent**
engaging in low level exercise such as walking: **23 per cent**

Average number of hours worked annually by Australian workers in 2008: **1,721**
Greece: **2,120**
Netherlands: **1,389**
USA: **1,792**
OECD average: **1,766**

Proportion of surveyed Australian men who believe that there is a "glass ceiling" for women in the workplace: **25 per cent**
of women: **81 per cent**

Proportion of surveyed Australian employers who say that staff development and retention is their leading goal: **39 per cent**
attracting more suitable staff: **23 per cent**
enhancing performance and productivity: **21 per cent**
other: **17 per cent**

Proportion of British workers who say they come up with new ideas at their desk: **10 per cent**
while drifting off to sleep: **30 percent**

Proportion of surveyed Australian workers who say they would quit their job if they won $1 million:
14 per cent
make a large donation to charity:
8 per cent

Proportion of Australian men on low incomes who are married:
37 per cent
high incomes: **70 per cent**

Doing nothing is very hard to do.
You never know when you're finished.

Leslie Nielsen, Canadian actor and comedian

***

Proportion of surveyed Australian workers who say they have been bullied:
40 per cent
proportion of these who named their boss as the culprit:
73 per cent

Proportion of Australian workers who use a car to get to work:
75 per cent
public transport: **12 per cent**
other: **13 per cent**

Proportion of surveyed mothers in the workforce in Australia who were back at work before their baby was 9 months old:
40 per cent
6 months old: **25 per cent**
3 months old: **10 per cent**

Estimated number of working days lost in Australia each year due to hangovers:
2.6 million

estimated annual cost:
$473 million

Estimated proportion of employees in Australia who are not entitled to either paid holidays or paid sick leave in their main job:
24 per cent

Proportion of surveyed workers in the US who say their boss blames others to cover up their mistakes:
23 per cent
fails to keep promises: **39 per cent**
has made negative comments about them to others in the previous year: **27 per cent**

Proportion of surveyed Australian workers who feel that body language is more revealing than spoken words in the office:
79 per cent
who feel they are usually correct in their assessment:
87 per cent

Proportion of surveyed Australian workers who worked overtime in the week before the survey: **92 per cent**
proportion of those who did not get paid for it: **79 per cent**
average extra time worked: **10.5 hours**

Proportion of employees in Australia who were trade union members in relation to their main job in August 2008:
19 per cent
proportion of public sector employees: **42 per cent**
private sector employees: **14 per cent**

Proportion of surveyed US workers who feel they are more productive when they work alone: **35 per cent**
in a small group: **37 per cent**
large group: **5 per cent**
other: **23 per cent**

Proportion of surveyed Australians who believe that their occupation might make or break a successful date:
48 per cent

The trouble with unemployment is that the minute you wake up in the morning you're on the job.

Slappy White, US comedian and actor

***

Proportion of surveyed Australian workers who take their own lunch from home, that choose sandwiches or salads:
40 per cent

Age of surveyed Australian workers who have the least respect for their boss:
35 and over

Proportion of surveyed US workers who see religion as the most taboo topic to discuss in the workplace:
29 per cent
office gossip: **27 per cent**
their personal life: **16 per cent**
politics: **14 per cent**
money: **14 per cent**

Number of Australians who will retire during the next ten years:
over 4 million
average age at retirement: **57 years**

Average number of sick days taken by Australians each year:
8.5
cost per employee: **$2,500**

Proportion of surveyed working Australians who feel they would have a better working life if they had studied further:
71 per cent

Proportion of surveyed Australian workers who say they have been called into the office during their days off or holidays:
74 per cent

Proportion of surveyed Australian women who shave their legs once a week or less:
52 per cent
who have gone to work with unshaven legs:
85 per cent

Proportion of surveyed Australian workers in sales and clerical positions who say stress is a significant factor in their life:
87 per cent

Ages at which Australians are most likely to say that their life is out of control: **30–49 years**
for whom work causes the most stress:
18–29 years

Proportion of surveyed Australian workers who generally bring a homemade lunch to their workplace every day:
4 per cent

Proportion of surveyed full-time workers in Australia who didn't take their annual leave in 2009–10: **1 in 8**
proportion who don't use their 4 weeks annual leave each year:
60 per cent

Proportion of surveyed Australians who believe it is acceptable to use swear words in the workplace: **25 per cent**
Japanese: **13 per cent**

It's just a job. Grass grows, birds fly, waves pound the sand. I beat people up.

Muhammad Ali, former US boxer

***

Proportion of surveyed office workers in Australia who consider the workplace an ideal hunting ground for romance:
80 per cent

Proportion of surveyed US workers who feel their work is viewed the same way when they do it remotely rather than physically in the office:
55 per cent

Proportion of surveyed Hong Kong office workers who say they suffer from depression: **25 per cent**
anxiety: **29 per cent**
stress: **35 per cent**

Proportion of surveyed senior executives in the US who feel comfortable looking for a new job while still employed:
75 per cent

Proportion of surveyed Australian workers who say they are happy in their job: **29 per cent**
unhappy: **45 per cent**

neither happy or unhappy: **26 per cent**

Proportion of employed people in Australia who work some hours at home in either their first or second job:
24 per cent

Proportion of surveyed Australian workers who intend to look for a new job within the next 12 months:
59 per cent

Proportion of surveyed British women who said they had laddered their tights at work: **31 per cent**
men who had split their trousers: **14 per cent**
men who had forgotten to do up their zipper: **19 per cent**

Proportion of surveyed Australians who said their job makes them ill:
10 per cent

Proportion of surveyed Australian employees who turn up for work when they are sick partly because of fear of falling behind in their work duties: **61 per cent**
fear they will be seen as a 'slacker': **48 per cent**
fear they will be sacked: **45 per cent**

Proportion of surveyed Australians who blame their parents for their size: **22 per cent**

who believe their being fat seriously interferes with career advancement: **33 per cent**
who have dieted just to impress their boss: **4 per cent**

Reported average amount of unpaid overtime worked each day by full-time employees in Australia: **70 minutes**
total unpaid hours worked each year across Australia: **2.14 billion**
their worth: **$72 billion**

I only go to work on days that don't end in a 'y'.

Robert Paul, US political philosopher

***

Proportion of surveyed Australians who are happy with their career and don't want to change:
22 per cent
who hate their job:
9 per cent

Proportion of surveyed American workers who say they are loyal to their company:
57 per cent
who say their company is loyal to them:
25 per cent

Proportion of surveyed US workers who say there are not enough managers in their work place:
52 per cent

Proportion of surveyed Australian women who said they have had a crush on a work colleague:
32 per cent

Proportion of surveyed Australian workers who say that disruptive, talkative colleagues are their biggest distraction:
44 per cent

Australia's unemployment rate in May 2010:
5.2 per cent
number of people unemployed:
610,400

Proportion of Hong Kong workers who work more than 48 hours per week:
41 per cent

Average number of steps taken per day by surveyed nurses in Australia: **15,000**
waiters: **9,000**
builders: **1,183**

Proportion of chief executives in Australia who are women:
2 per cent
proportion of ASX200 boards that have no women at all:
50 per cent

Average office space per worker in Sydney's CBD in 2009:
18.57 square metres

It's a recession when your neighbour loses his job: it's a depression when you lose yours.

Harry S. Truman, former US President

***

Proportion of surveyed Australians who say they feel stressed:
87 per cent
proportion of these that blame their work:
75 per cent

Proportion of Australian families in which at least one parent
works at least some time after 7.00pm:
60 per cent

Number of registered beekeepers in NSW: **about 3,000**
in Victoria: **about 2,100**

Proportion of surveyed Australians who feel that butchers are happy in their jobs:
30 per cent (the highest)
service station attendants, bank tellers and sandwich hands:
10 per cent (the lowest)

Proportion of surveyed professional workers in Australian who say they laugh and have fun at work:
25 per cent

Proportion of typing that is done by the left hand:
56 per cent

APPEARANCES

My wife was afraid of the dark ... then she saw me naked and now she's afraid of the light.

Rodney Dangerfield, US comedian

Guys: No Shirt, No Service – Gals: No Shirt, No Charge

Author unknown

***

Proportion of surveyed Australians who feel advertising is too focussed on images promoting youth:
83 per cent

Proportion of surveyed 15–64 year-old Australian women who refuse to participate in activities because they feel too vulnerable and exposed:
67 per cent

Proportion of surveyed Australian women who say that beauty advertisements make them feel bad about the way they look:
49 per cent

Estimated number of frowns required to cause one brow wrinkle:
about 200,000

Amount of time that surveyed urban Chinese men aged 18–60 reportedly spend each day gazing in the mirror:
8.6 minutes
in Shanghai:

17 minutes

Proportion of surveyed French people who weighed themselves daily: **15 per cent**
Americans: **12 per cent**

Proportion of surveyed Americans who believe most Americans are overweight:
90 per cent
who believe that they themselves are overweight:
39 per cent

Proportion of surveyed Australian women who consider themselves overweight:
75 per cent
who would be happier if they were a few kilograms lighter:
85 per cent
who would like to change something about their body:
95 per cent

Proportion of Australian women who want full-figured women in advertisements:
72 per cent
less airbrushing in photographs:
79 per cent

Average amount that Western men are taller than Western women:
11–12 centimetres

Proportion of surveyed 10–14 year-old Australian girls who think famous girls on television and in magazines look beautiful:
86 per cent

The average woman would rather have beauty than brains, because the average man can see better than he can think.

Anonymous

***

Proportion of Australian women who don't like the shape of their body:
67 per cent
who dislike it intensely:
10 per cent

Proportion of surveyed 18–64 year-old Australian women who are very satisfied with their looks:
6 per cent

Proportion of surveyed Australian women who would date a man who wears make-up:
30 per cent

Age at which surveyed British women feel that women peak: **39**
at which their looks start to decline: **46**

Proportion of surveyed Australian men who think that brunettes are more attractive than blondes or redheads:
51 per cent

that brunettes are more intelligent:
81 per cent

Proportion of surveyed Australians who say they are satisfied with their partner's appearance:
94 per cent
who feel that their own looks make them seem older than they actually are:
80 per cent

Proportion of surveyed Australian men who say they immediately notice if their wife or girlfriend has a new haircut or new outfit: **56 per cent**
who eventually notice: **38 per cent**
never notice: **6 per cent**

The lovely thing about being 40 is that you can appreciate 25-year old men more.

Colleen McCullough, Australian author

***

Proportion of surveyed American husbands who always tell their wife she looks beautiful: **15 per cent**
often: **42 per cent**
sometimes: **37 per cent**
rarely or never: **6 per cent**

Estimated number of times an American woman shaves her legs in her lifetime: **6,626**
total hours spent doing so: **533**

According to surveyed Australian men, the most attractive height for a woman:
163 centimetres
waist: **76 centimetres**
hips: **102 centimetres**
equivalent size: **14**

Proportion of surveyed British men who couldn't recall their partner's eye colour: **12 per cent**
didn't know her date of birth: **10 per cent**
didn't know her natural hair colour: **8 per cent**
didn't know her job title: **11 per cent**
couldn't name her best friend: **12 per cent**

Proportion of surveyed American males who have a moustache:
22 per cent

Proportion of surveyed Australian female users of cosmetics in Australia who apply make-up before going shopping: **50 per cent**
to the beach: **32 per cent**
to the gym: **24 per cent**

Proportion of surveyed men who say that a woman's face is much more important than her breasts, legs or figure:
68 per cent

Proportion of surveyed Australian men who look for great breasts in a woman: **8 per cent**
nice legs: **16 per cent**
a fat-free figure: **16 per cent**

Proportion of surveyed Australian men who viewed a size 16 model as normal looking:
58 per cent
really hot: **16 per cent**
overweight: **26 per cent**

Number of plastic surgery procedures performed in Argentina:
in 2008: **about 50,000**
in 2003: **31,000**

I'd love to kiss you, but I just washed my hair.

Bette Davis, US actress (Cabin in the Cotton, 1932)

***

While on holiday, the average time UK women aged 16-60 spend each morning finding an acceptable outfit to wear:
10 minutes
selecting evening clothes at night:
10 minutes

Average time taken by British women each weekday morning to find an acceptable outfit to wear:
16 minutes

Number of people per 100,000 in Australia who have had plastic surgery: **66**
world ranking: **7th highest**

Year the first Miss World took place, in London, won by Kiki Haakonson of Sweden: **1951**

Number of teenage girls in Italy who had plastic surgery in 2008: **10,000**

Proportion of surveyed NSW residents who had measured their waist in the previous six months: **40 per cent**
who had tried to reduce their waist size: **55 per cent**

Proportion of surveyed British women who have trouble storing their shoes: **33 per cent**
who have more than thirty pairs of shoes in their closet:
17 per cent

Proportion of surveyed Australians who say they have thought about having plastic surgery:
61 per cent
who have actually undergone cosmetic surgery or a non-surgical procedure:
10 per cent

Proportion of surveyed Australian men who would lie if their partner said 'does my bum look big in this?':
52 per cent

Number of outfits UK women try on each morning before deciding what to wear: **2**
average time spent deciding what to wear before going out on a weekend: **20 minutes**

Proportion of surveyed men who say that women spend far too much time worrying about their appearance:
67 per cent

BIRTHS AND DEATHS

My luck is so bad that if I bought a cemetery, people would stop dying

Ed Furgol, US professional golfer

If you die in an elevator, be sure to push the Up button.

Sam Leverson, US humourist, writer, journalist and television host

***

Estimated time a person could breathe in an average size coffin:
about 1 hour

Number of infant deaths per 1,000 live births in Australia:
in 2009: **4.7**
in 1994: **9.2**

Proportion of births in Australia that are to married women:
66.5 per cent

Proportion of babies in Australia that are conceived through IVF: **3 per cent**

Age group of women who give birth to the most babies in Australia: **30–34**
number of births per 1,000 of these women in 2008: **127.8**
in the age group 25–29: **105.8**
in the age group 20–24: **53**

Proportion of surveyed Australian women who were unaware that they can become pregnant on only two days each month:
94 per cent

Proportion of suicides in Australia in 2008 that were by:
hanging: **53 per cent**
poisoning by drugs: **12 per cent**
alcohol and motor vehicle exhaust: **11 per cent**
firearms: **7.8 per cent**

Proportion of births in Britain that are by caesarean section:
23 per cent
in Australia: **27 per cent**

Proportion of babies born out of wedlock in Australia in 2008:
33.5 per cent

Median age of new mothers in Australia in 2008:
30.7 years
fathers: **33.1 years**

Proportion of pregnancies for women aged under 40 years in Australia that end in a miscarriage:
20 per cent

Number of births registered in Australia:

in 2008 (the largest on record): **296,610**
in 2005: **263,409**

Proportion of surveyed new mothers in Australia who consumed alcohol during pregnancy and after while breastfeeding:
47 per cent

Death is nature's way of telling you to slow down.

Terry Pratchett, British novelist (Strata, 1981)

***

Estimated proportion of surveyed pregnant women in Australia in 2008 who smoked sometime during their pregnancy:
10–20 per cent

Factor by which a woman's uterus expands when she is pregnant:
up to 500 times

Estimated number of people who die in traffic accidents worldwide each year:
about 1.3 million
estimated number in 2030:
2.4 million

Estimated number of births in the world in 2009:
139 million
deaths:
56 million

Number of cot deaths in Australia in 2009:
about 90

Estimated proportion of deaths in the world that go undocumented:
about 2 in 3

Proportion of all deaths of those aged 15–24 in Britain that are suicide:
23 per cent

Estimated number of Australians alive today who will die prematurely from heart problems caused by obesity:
123,000

Proportion of births that are naturally-conceived identical triplets:
1 in 200 million

Reported number of sons fathered by Ramses II, Pharaoh of Egypt: **44–56**
daughters: **40–44**
year he died: **1213BC**

Number of registered male deaths in Australia 2008:
73,500
female: **70,400**

Reported number of reported deaths when a Senegalese ferry capsized in the Atlantic Ocean near the Gambria Sea:
1,863

year: **2002**

Estimated number of people worldwide who die each year from too much exposure to the sun:
about 60,000

I always wait for the Times each morning. I look at the obituary column and, if I'm not in it, I go to work.

A.E. Matthews, British actor (quoted in Filmgoer's Book of Quotes, 1973)

***

Number of births per 1,000 Americans in 2006: **14.3**
deaths: **8.1**
marriages: **7.3**
divorces: **3.6**

Number of infant deaths in Australia:
in 2008: **1,200**

in 2006: **1,260**

Estimated proportion of humans that have ever lived that are dead:
90 per cent

Number of deaths from an eruption of the Tambora volcano, on Sumbawa Island, Indonesia in 1815:
over 71,000

Estimated number of deaths resulting from the 1976 earthquake in Tangshan, China:
about 250,000
the 2008 Sichuan earthquake in China:
about 70,000

Proportion of surveyed Australians who believe in:
life after death: **53 per cent**
Heaven: **56 per cent**
Hell: **38 per cent**
the Devil: **37 per cent**
witches: **22 per cent**

Proportion of births in the UK in 2008 that were to non-UK born mothers:
14 per cent

Amount left in a Sydney woman's will in 2009 to animal welfare groups:
$3.27 million

number of animals she reportedly owned: **0**

Everybody wants to go to heaven; but nobody wants to die.

Anonymous

***

Proportion of surveyed Americans who would like to live to age 100: **63 per cent**
who would not like to: **32 per cent**
who are unsure: **5 per cent**
who expect to do so anyway: **8 per cent**

Life expectancy for a male in Australia currently aged 65 years:
16.1 years
a female: **21.0 years**

Proportion of all births in Australia in 2008 where the paternity was not acknowledged:
3 per cent
of ex-nuptial births: **9 per cent**

Number of cremations in Australia each year:
about 65,000

Proportion of babies born in Australia in 2008 who did not survive until their first birthday:
1 in 240
in 1901–1910: **1 in 12**

Proportion of newborn deaths in Australia in 2008 that occurred on the first day of life:
39 per cent
in the remainder of the first four weeks of life:
31 per cent

Proportion of registered male deaths in 2008 where he was married at the time of his death:
54 per cent
widowed:
19 per cent
never married:
14 per cent

Proportion of registered female deaths in 2008 where she was married at the time of her death:
26 per cent;
widowed:
57 per cent
never married:
8 per cent

Number of suicides in Australia in 2008: **2,191**
proportion who were male: **78 per cent**

Number of deaths from drowning in Australia in the year ending:
30 June 2009: **312**
30 June 2008: **271**

Somewhere on this globe, every ten seconds, there is a woman giving birth to a child. She must be found and stopped.

Sam Leverson, US humourist, writer, journalist and television host

***

Number of confinements resulting in multiple births in Australia in 2008: **4,600**
in 1988: **3,700**

World fertility rate of developed countries in 2009: **2.6**
in least developed countries: **4.6**

Number of caesareans performed in the USA in 2007 (the highest on record):
1.4 million
in Australia:
about 100,000

Proportion of males born in Australian between 1901 and 1910 who did not survive to age 40:
24 per cent
females: **22 per cent**

Number of drownings in Australia in 2008-9 that were children aged under 5 years: **32**

proportion of these drownings that occurred in swimming pools:
60 per cent

Birth rate in 2005 of foetuses diagnosed with Down Syndrome:
35 per cent
in 1995: **81 per cent**

Reported number of executions in the world in 2008: **2,390**
number of these in China: **1,718**
in Saudi Arabia: **102**
in USA: **37**
in Pakistan: **36**
in Iraq: **34**

Proportion of all births in Australian in 2007 that were caesarean: **30.9 per cent**
for first time mothers: **32.0 per cent**
for women who had previously given birth by caesarean section: **83.0 per cent**

Proportion of surveyed Australian mothers who say they:
are happy: **68 per cent**
miss their freedom: **52 per cent**
are now more like friends than lovers with their partner:
21 per cent

Reported number of Australian women on the pill:
1.4 million
as a percentage of all women of reproductive age:
30 per cent

He is one of those people who would be enormously improved by death.

H.H. Munro, (aka Saki), British writer

***

Number of births in Britain in 2008:
791,000
fertility rate (the highest since 1973):
1.97

Years of the bubonic plague (The Black Death):
1348–51
number of people who died from it:
between 75 and 200 million

Median age of mothers of ex-nuptial births in 2008:
27.0 years

Median duration of marriage before the birth of the first child
in 2008: **4.1 years**
in 1998: **4.6 years**

Year the rack was introduced to the Tower of London (by the Duke of Exeter):
about 1420

MARRIAGE AND DIVORCE

I've been in love with the same woman for forty-one years. If my wife finds out, she'll kill me.

Henry Youngman, British-born comedian

Funny thing how you first meet the woman that you marry. I first met the wife in a tunnel of love. She was digging it.

Les Dawson, British comedian

***

Proportion of Australian couples affected by infertility:
17 per cent

Median age of first time brides in Australia:
in 2008: **27.7 years**
in 1989: **24.2 years**

Median age of first time grooms in Australia:
in 2008: **29.6 years**
in 1989: **26.3 years**

Median time for Australian men to remarry after divorce:
3.5 years
women: **4.2 years**

Ratio of widows to widowers in the UK:
3 to 1

Proportion of teenage mothers in the USA today who are unmarried: **80 per cent**
in 1950: **13 per cent**

Estimated number of men in Britain who live legally with multiple wives:
about 1,000

Proportion of divorced couples in Australia who separated:
in their first year of marriage: **6 per cent**
within the first five years: **33 per cent**
between 5 and 9 years: **22 per cent**

Proportion of Australians aged over 15 years who are married:
49.6 per cent
in 1996: **53.3 per cent**

Age group at which women are most likely to accidentally fall pregnant: **25–29**
proportion of surveyed women with an unplanned pregnancy who kept the baby: **56 per cent**
had an abortion: **29 per cent**
miscarried: **13 per cent**
offered the baby for adoption: **2 per cent**

Proportion of men and women divorcing in the UK in 2005 that had a previous marriage ending in divorce:
20 per cent
in 1981: **10 per cent**

Estimated proportion of Australian women aged 30–34 who have a degree that are married:

81 per cent
without a degree: **53 per cent**

Age of Anna Nicole Smith when she married
J. Howard Marshall II in 1994: **26**
his age: **89**

It's better to have loved and lost than to do forty pounds of laundry a week.

Laurence J. Peter, Canadian educator

***

Proportion of surveyed fathers in Hong Kong who say they find parenting so tough that they would not repeat the experience:
25 per cent

Estimated proportion of marriages that survive infidelity:
about 70 per cent

Estimated proportion of single Australian men who will never marry: **1 in 3**
of single women: **1 in 4**

Proportion of surveyed teenage mothers in Queensland who smoked throughout their pregnancy:
44 per cent

Median age at marriage for all grooms in Australia in 2008:
31.6 years
brides: **29.3 years**

Number of divorces in Australia:

in 2008: **47,209**
in 2004: **52,747**
proportion that involved children: **50 per cent**

Proportion of surveyed Australian women who believe that pregnancy would affect their careers:
50 per cent

Proportion of surveyed Australian married couples who had split in the previous 5 years who said their social life had subsequently improved:
46 per cent
who said it cost more than $100,000 to start a new life:
19 per cent

Estimated proportion of marriages in Australia today that are likely to end in divorce:
32 per cent

Proportion of marriages in Australia where the groom is older than the bride:
66 per cent

Number of weddings in the USA in 2009:
2,272,624
average cost of a US wedding: **$US30,860**
average number of guests: **169**
average length of an engagement: **17 months**

Wedding anniversary at which a gift of iron should be given:
6 years
coral or jade: **35 years**

Proportion of Australian couples who live together before marriage:
75 per cent
or those who married in the 1960s (and are still married):
3 per cent

In Hollywood a marriage is a success if it outlasts milk.

Rita Rudner, US comedienne, writer and actress

***

Proportion of lone parent families in the UK that are headed by a woman:
90 per cent
proportion of lone mothers who are single (never married):
50 per cent

Proportion of de facto couples in Australia who are of the same gender: **about 4 per cent**
in NSW: **5.4 per cent**
Victoria: **4.5 per cent**
Tasmania: **2.6 per cent**

Number of divorces in Australia in 2008 (a 19-year low):
47,209
median age of men at divorce: **44.1 years**
of women: **41.4 years**

Number of divorces in the European Union in 2007:
about 1 million
number that included couples from different countries:
about 140,000

Proportion of adult American men who:
are currently married: **60 per cent**
are divorced or separated: **10 per cent**
never married: **27 per cent**
other: **3 per cent**

Proportion of Australians who use a civil celebrant to officiate their marriage:
65 per cent

Median age at separation for men in Australia in 2008:
40.5 years
women: **37.8 years**

Number of registered marriages in Australia:
in 2008: **118,756**
in 2004: **110,958**
in 1988: **116,816**
in 1984: **108,655**

Median duration of the marriage for divorces granted in Australia in 2007: **12.5 years**
in 1998: **11.2 years**

Number of women aged 16–19 in Australia in 2008 who married for the first time: **2,020**
in 1989: **8,058**

Number of women aged 75 and over in Australia who married in 2008: **157** men: **449**

Proportion of marriages in Australia in 2008 that were in November (highest month):
12.7 per cent
July (lowest month):
4.2 per cent

Whenever you want to marry someone, go have lunch with his ex-wife.

Shelley Winters, US actress

***

Number of marriages in Australia:
in November 2008 (highest month): **15,109**
in July (lowest month): **5,008**

Number of marriages in Australia that took place on 29 February 2008: **564**

Most popular date to get married in Australia in 2008:
4th October
number of weddings on that day: **2,221**

Date that had the least weddings in Australia in 2008:
30th January
number of weddings on that day: **31**

For Australian men who divorce the most likely age at which they married: **27 years**
for women: **24 years**
ages at which they most likely divorce:
35–39 years

Proportion of all marriages in the US in 2008 that were between a man who was 10-14 years older than his wife:
5 per cent
6-9 years older:
11.6 per cent

Proportion of all marriages in the US in 2008 that were between a woman who was 10-14 years older than her husband:
1 per cent
6-9 years older:
2.6 per cent

Proportion of couples married in Australia in 2008 who were both doing so for the first time:
70 per cent

Of the divorces in Australia in 2008, proportion that were filed by the wife alone:
37.3 per cent
husband alone: **27.5 per cent**
jointly: **35.2 per cent**

Proportion of divorces in Australia in 2008 that were to marriages of five years or less:
22.5 per cent
25 years or more: **17.2 per cent**

Proportion of divorces in Australia in 2008 where the marriage was at least 20 years:
18.6 per cent
at least 30 years: **4.7 per cent**

Proportion of all marriages in the US in 2008 that involved a remarriage for one or more of the spouses:
46 per cent

BABIES

Have you noticed that all the people in favour of birth control are already born?

Benny Hill, British comedian

Having a baby changes the way you view your in-laws. I love it when they come to visit now. They can hold the baby and I can go out.

Matthew Broderick, US actor (on Sarah-Jessica Parker's in-laws. Quoted in Sunday Mail, 26 Jan 2003)

***

Average number of labour pains for first time mothers: **about 135**

Average sleep time per 24 hours for:
babies: **16 hours**
12-year-olds: **8 hours**
adults: **7 hours**

Age at which most babies can recognise their own name:
4–7 months

Proportion of surveyed new mothers in Australia who were having sex again within 8 weeks after the birth of their baby:
80 per cent
within 10 weeks:
98 per cent

Average survival rate for a baby born at:

22 weeks: **4.6 per cent**
23 weeks: **46 per cent**
24 weeks: **59 per cent**
25 weeks: **82 per cent**

Weight of the tiniest baby ever to survive (Rumaisa Rahman, Miami): **244 grams**
weeks premature: **14**
year: **2004**

Number of recorded twins born to different fathers: **about 10**

Proportion of babies in Sierra Leone who die before their first birthday:
15–16 per cent

Estimated proportion of three-month-old babies in the US who are regular television viewers:
40 per cent

A good film is when the price of the dinner, the theatre admission and the babysitter were worth it.

Alfred Hitchcock, British filmmaker and producer

***

Proportion of babies affected by infantile colic in their first six months: **28 per cent**
age at which symptoms usually peak: **6 weeks**

Average number of disposable nappies thrown away in Australia each year:
up to 800 million

Number of babies born in the USA in 2007 (the highest in 50 years): **4,315,000**
in 2008: **4,247,000**

Amount of milk produced each day by a lactating mother:
about 1.7 litres

Reported success rates for women using IVF
at age 31: **27 per cent**
at 40: **10.5 per cent**
at 43: **2.6 per cent**
over 45: **0.7 per cent**

Average amount of time twins are born before single births:
19 days

Number of treatment embryos reportedly in storage in Australia in 2006: **118,700**
in 1994: **about 22,000**

Number of months before you are born that your teeth start growing: **6**

The most foetuses found in a woman (in the womb of an Italian housewife):
15 (10 girls and 5 boys)
year: **1971**

Estimated percentage of childbirth labours worldwide that have life-threatening complications:
15 per cent

Amount of time after birth that an infant can breathe and swallow simultaneously:
about 6 months

Number of services provided to women in Australia under the IVF Assisted Reproduction Services scheme in 2007–8:
55,055

Number of babies born in Australia each year with a congenital heart disease: **about 100**

proportion of heart defects that can be fully repaired in the first year of life: **about 80 per cent**

When I was born I was so surprised I didn't talk for a year and a half.

Gracie Allen, US comedienne

***

Proportion of surveyed Australians who believe that women get post-natal depression because they have unrealistic expectations of motherhood: **57 per cent**
who feel that post-natal depression will go away on its own without treatment: **25 per cent**

Proportion of abortions in the UK that are carried out before 13 weeks:
87 per cent

Proportion of surveyed Australian mothers aged 18–45 who said they drank alcohol throughout their last pregnancy:
34 per cent
who said they will likely do so through their next pregnancy:
24 per cent

Proportion of surveyed Australian mothers aged 18–45 who said they smoked throughout their last pregnancy:
16 per cent

who said they will likely do so through their next pregnancy:
4 per cent

LAW AND ORDER

I tell you, I come from a tough neighbourhood. Why, just last week some guy pulled a knife on me. I could see it wasn't a real professional job. There was butter on it.

Rodney Dangerfield, US comedian

When a naked man is chasing a woman through an alley with a butcher knife and a hard-on, I figure he isn't out collecting for the Red Cross.

Clint Eastwood, US actor and director (Dirty Harry, 1971)

***

Number of times former cult leader Charles Manson has been denied parole since 1978: **11**
most recent time: **May 2007**
year he is next eligible to apply: **2012**
his inmate number at Corcoran State Prison: **B33920**

Number of cases before Magistrates' courts in Australia in 2008 that were for traffic offences:
264,800
as a proportion of all offences:
about 50 per cent

Proportion of illicit drug offences proven guilty in Australian magistrate's courts:
99 per cent
higher courts: **97 per cent**

Homicide rate (per 100,000 people) in the USA:
in 2008: **5.4**
in 1975: **9.6**
in 1963: **4.6**

Proportion of serial killers in the USA who are male:
88 per cent
their average age: **28.5 years**

Proportion of women in US prisons or goals who are single mothers:
77 per cent

Number of US presidents: **43**
number whose previous occupation was a lawyer: **25**

Proportion of prisoners in Britain who are a foreign national:
15 per cent
in the USA: **20 per cent**

Proportion of state prisoners in the USA who have prior service in the US armed forces:

10 per cent
in 1986: **20 per cent**

My husband and I didn't sign a pre-nuptial agreement. We signed a mutual suicide pact.

Roseanne Barr, US actress, comedienne, writer, television producer and director

***

Proportion of surveyed Britons who personally knew someone who had been murdered:
1 in 3

Number of suspects in the board game *Cluedo:* **6**
murder weapons: **6**
rooms: **9**
permutations: **324**

Estimated amount of money that is illegally laundered in Australia each year:
about $4.5 billion

Number of Nazis sentenced to death at the 1946 Nuremberg Trials: **12**
to life imprisonment: **1 (Rudolf Hess)**
jailed: **5**
acquitted: **3**

Number of judges in the original High Court of Australia: **3**
year: **1903**

Reported average IQ of serial killers in the USA: **107**
the general US population: **100**

Number of prison inmates in Australia as at 30 June 2009:
29,317
proportion who were female: **7 per cent**
the median age of all prisoners: **33.4 years**
proportion who had served time previously: **56 per cent**

Proportion of prison inmates in Australia who had served a sentence in an adult prison prior to their current incarceration:
59 per cent
median sentence length: **3 years**

Number of murders in New York City:
in 2008: **523**
in 2006: **579**
in 1990: **2,245**

Estimated number of people tried in China during 2006:
about 593,000
number who were found innocent: **1,464**

Date the first man was hanged in the colony of New South Wales (Thomas Barrett, for stealing food):
27 February 1788

Proportion of surveyed Americans who feel that DNA testing is very reliable for identifying an individual: **68 per cent**
somewhat reliable: **23 per cent**
unreliable: **7 per cent**

You don't know a woman till you've met her in court.

Norman Mailer, US novelist, journalist, essayist, poet, playwright and screenwriter

***

Estimated proportion of Australians who die without making a will:
60 per cent

Number of justices of the Australian High Court: **7**
number who are women:
3 (Bell, Crennan and Kiefel)

Estimated number of Australians affected by identity theft in 2008: **4.4 million**
total cost: **$3.5 billion**

Proportion of surveyed Americans in 2009 who felt that gun laws should be stricter:
44 per cent
in 1990: **78 per cent**

Proportion of surveyed Australian women drivers who say they do not drink and drive:
66 per cent
of men: **47 per cent**

Number of hospital admissions in Britain in the 12 months ending November 2008 for wounds due to assault by a knife or other sharp object: **5,350**
in 2007: **13,795**

Proportion of international child abductions for which the mothers are responsible:
70 per cent
proportion of these children who were aged younger than nine years:
78 per cent

Proportion of surveyed Australians who say they have a lot of confidence in the criminal justice system:
35 per cent
who have confidence in the courts to consider the rights of the accused: **70 per cent**
the victim: **50 per cent**

Number of justices on the US Supreme Court: **9**

Number of people killed or wounded by guns in the US each day: **276**
of those adults shot, the number who die: **75**
of children: **9**

Estimated proportion of Australians who were a victim of credit card fraud or computer hackers:
20 per cent

Prison sentence (upheld on appeal) given to a US man in 2009 convicted for multiple counts of sexual abuse:
4,060 years
year he will be eligible for parole:
3209

What do you call 500 lawyers lying on the bottom of the Ocean? A good start.

Danny DeVito, US actor (The War of the Roses, 1989)

***

Number of Australians in prison overseas on 30 June 2009:
in the USA: **36**
in New Zealand: **30**
in Vietnam: **25**
in Britain: **24**
in China: **19**
in Indonesia: **19**

Year Al Capone was born: **1899**
year he started his prison sentence for tax evasion: **1932**
year he died: **1947**

Number of geckos a German tourist leaving New Zealand in 2009 reportedly had hidden in his underwear: **23**
number of skinks: **20**

Number of people in Australian jails on 30 June 2009:
29,300

Proportion of murder cases in Australia in 2008 that were solved within 30 days of the body being found:
74 per cent
in NSW: **64 per cent**
in Victoria: **59 per cent**

Estimated number of Australians who have been exposed to some type of scam or fraud in the past 12 months:
about 6 million
total amount that they were fleeced:
almost $1 billion
estimated proportion of frauds that are reported to police:
25 per cent

Proportion of young people sentenced to detention in NSW in 2008 who were male:
92.8 per cent
female: **7.2 per cent**

Reported proportion of Americans who have spent some time in prison:
1 in 37

The number of people in the USA in 2008 who were either in jail or prison, on probation or on parole:
7.3 million
as a proportion of the US adult population:
3.2 per cent

Age group in Australia most likely to be assaulted:
15–24 years
least likely: **44–64 years**

PEOPLE AND PLACES

As you get older three things happen. The first is your memory goes and I can't remember the other two.

Sir Norman Wisdom, British comedian and actor

Give a man a fish and he eats for a day. Teach him how to fish and you get rid of him all weekend.

Zenna Schaffer

***

Estimated number of books sold by author Enid Blyton by 2010:
over 600 million
number of languages: **90**

Proportion of surveyed Americans who attend religious ceremonies at least occasionally:
90 per cent
who believe in God: **76 per cent**
who believe in an afterlife: **60 per cent**

World record for 100 metres running for men aged 95–99 (Kozo Haraguchi, Japan, 2005):
22.04 seconds
age he began competing in track and field events:
65 years

Combined weight of two minibuses pulled by Georgian wrestler Lasha Pataraia's left ear:
4.5 tonnes
distance: **41.1 metres**
year: **2006**

Amount won by an 84-year-old grandmother (Josephine Crawford) on the spin of a 5-cent poker machine in Atlantic City in April 2006:
$A13.4 million

Proportion of surveyed Americans who could identify a picture of Harry Potter:
57 per cent
Tony Blair, when he was prime minister of the UK:
50 per cent

Record number of bratwursts eaten in ten minutes (by Takeru Kobayashi, Japan, 2006): **58**
amount of fat consumed: **1.45 kilograms**

Total weight of mud reportedly eaten by a Mongolian girl since 1994 because she likes the taste:
1.55 tonnes

Number of times that Glynn "Scotty" Wolfe, USA, was divorced (a world record): **28**
his longest marriage: **7 years**
shortest: **19 days**
reported value of his estate on his death in 1997:
$US337
his age: **88 years**

Number of US states named after a US President:
1 (Washington)

Longest solo drumming kit marathon (by Australian Suresh Joachim): **84 hours**
year: **2004**

Longest natural eyelash (Daniel Williams, USA): **about 5 centimetres**

Some folks are wise and some otherwise.

Josh Billings, US humourist and lecturer

***

Fastest time to eat 38 soft-boiled eggs (Peter Dowdeswell, UK):
1 minute 15 seconds
year: **1984**

Youngest person to become a great-great-great grandmother (Harriet Holmes, Canada):
88 years 50 days
year: **1987**

Number of labours of Hercules:
12

Number of days US President John F. Kennedy spent in office:
1,036
year he was born: **1917**
year he became President: **1961**

Amount paid at a US auction in April 2007 for a pair of John Lennon's sunglasses: **$US30,000**
Bill Clinton's saxophone: **$US54,000**

Length of the longest ever speech (given by Jayasimha Ravirala, April 2007, on the topic of 'Personality Development Concepts'):
111 hours

Fastest time to pluck a turkey (Vincent Pilkington, Ireland):
1 minute 30 seconds
year: **1980**

Amount awarded in Arts prize money to painter Gustav Metzer whose work had previously been thrown away by a cleaner at the Tait Gallery in the UK:
£30,000
year: **2006**

Amount paid at a Christie's auction in New York in November 2006 for Andy Warhol's portrait of Mao Zedong:
$US17.36 million
its dimensions:
205 centimetresx155 centimetres
year it was created: **1972**

Year that Sir Robert Menzies died: **1978**
his age: **83 years**

Fastest time to eat a 30.5 centimetre pizza (Tom Waes, Belgium):
2 minutes 19.91 seconds

year: **2006**

Longest human chain (holding hands, stretching from Teknaf to Tentulia, Bangladesh):
about 1,050 kilometres
number of people:
about 5 million

Fastest time to run 100 metres barefoot on ice (Nico Surings, Netherlands):
17.35 seconds
year: **2006**

Knowledge speaks, but wisdom listens.

Jimi Hendrix, US singer and songwriter

***

Weight of the largest carved sapphire (Richard Sipe Sr, North Carolina, USA):
16.1 kilograms
number of carats: **80,500**
year: **2005**

Dimensions of the world's largest playable accordion (made by Giancarlo Francenella, Ancona, Italy):
253centimetresx190centimetresx85centimetres
year completed: **2001**

Fastest time to carve a face into a pumpkin (Stephen Clarke, USA):
24.03 seconds
year: **2006**

Longest time juggling a regulation soccer ball non-stop using only feet, legs and head without the ball ever touching the ground (by Martinho Eduardo Orige, Brazil):
19 hours 30 minutes
year: **2003**

Weight of the heaviest lemon

(grown by Aharon Shemoel, Israel):
5.265 kilograms
year: **2003**

Average speed of the winner of Indianapolis 500 in 2010 (Dario Franchitti):
260 kilometres per hour
in 1911 (Ray Harriun):
120.06 kilometres per hour

Amount reportedly paid for Jackson Pollock's painting "No. 5, 1948"(a record):
about $US140 million
the dimensions:
122 centimetresx244 centimetres
year of sale: **2006**

Smallest hand-made chess set (by Jaspal Singh Kalsi, India):
3.2 centimetresx3.2 centimetres
year: **2003**

World record for the 500 metre speed skating sprint (by Jeremy Wotherspoon, Canada):
34.03 seconds
year: **2007**
his age: **31 years**

Age of Australia's youngest Prime Minister (Chris Watson): **37**

year: **1904**

Most recorded leaves on a clover stem (found by Shigeo Obara, Japan): **18**
year: **2002**

I was born in a small town: Babylon, Long Island. And I'll tell you one thing about Babylon. The population never changes. Every time a kid is born, some guy leaves town.

Rodney Dangerfield, US comedian and actor

***

Height of world's tallest person (Leonid Stadnyk, Ukraine):
2.57 metres
weight: **about 200 kilograms**
year of birth: **1971**

Generally accepted number of heads of Hydra in Greek mythology: **9**

Estimated direct male descendants of Genghis Khan:
about 16 million
year he lived: **c 1162–1227**

Number of town and cities visited by Queen Elizabeth on her 1953–4 trip to New Zealand: **46**
number of separate functions she attended: **110**

Number of US Presidents who died on a 4 July:
3 (John Adams, Thomas Jefferson, James Monroe)

Most books typed backwards (by computer, Michele Santelia, Italy): **64**
number of words: **3,361,851**
number of characters: **19,549,382**
year of completion: **2007**

Highest speed while riding a motorcycle unsighted (Billy Baxter, UK):
265.33 kilometres per hour
size of the engine: **1,200 cc**

Longest leg hair (Wesley Pemberton, USA):
16.51 centimetres
year measured: **2008**

Longest distance jumped on a pogo stick (Ashrita Furman, USA):
37.18 kilometres
year: **1997**

Most combined fingers and toes on a living person (held by two people, Pranamya Menaria and Devendra Harne, India): **25**
number of fingers: **12**
number of toes: **13**

Number of kangaroos depicted on the Australian one dollar coin: **5**

Number of people who have won a Nobel Prize in two different fields:
2 (Marie Curie for Physics in 1903 and Chemistry in 1911,
Linus Pauling for Chemistry in 1954 and Peace in 1962)

Amount paid at a Melbourne auction for Sir Donald Bradman's first Test cricket bat (signed by the entire 1928–29 Australian team):
$145,000

When a man talks dirty to a woman, it's sexual harassment. When a woman talks dirty to a man, it's $3.95 a minute.

Anonymous

***

Longest time singing without stopping (Kim Sun-Ok, South Korea):
76 hours
number of tunes she sung: **1,283**
her age: **54**
year: **2009**

Most gallstones recovered from a patient (from K. Whittle Martin, UK, 1987):
23,530

Weight of the heaviest kidney stone removed (from Wazir Muhammad, Pakistan):
620 grams
year: **2008**

Combined length of the world's longest toenails (Louise Hollis, USA):
221 centimetres
year they were last cut: **1982**

Height of the world's tallest female model ('Eve', from the USA):
205 centimetres

Years that Ivan the Terrible lived: **1530–84**
number of times he married: **8**

Year Sir Walter Raleigh was beheaded: **1618**
his age: **about 66 years**

Years that Charles I lived:
1600–1649

Number of times Queen Isabella I of Castile reportedly bathed in her life:
2 (at birth and before her marriage)
years she lived: **1451–1504**

Proportion of people in New York who were observed to wash their hands after using a public toilet:
49 per cent

Years that Guy Fawkes lived:
1570–1606

Proportion of the world's population accounted for by China, India, Brazil and Russia:
42%

Number of people who visited Sydney:
in 2008: **25.7 million**
in 1999: **32.7 million**

Always remember that you are absolutely unique. Just like everyone else.

Margaret Mead, US cultural anthropologist

***

Area of the Netherlands:
41,526 square kilometres
proportion that is below sea level:
20–27 per cent

World's largest waist (Walter Hudson, USA):
302 centimetres
his weight: **545 kilograms**
year he died: **1991**

World's widest tongue (Jay Sloot, Australia):
7.9 centimetres
longest tongue (Stephen Taylor, UK):
9.8 centimetres

Longest distance sailed single-handed in 24 hours (Thomas Colville of France):
1,163.9 kilometres
year: **2008**

Number of museums and art gallery locations in Australia in June 2008: **1,456**
number of employees: **7,856**

Number of living languages in the world: **about 6,800**
proportion that are limited to a single country: **83 per cent**

Number of people worldwide that speak, as a first language:
Mandarin: **885 million**
Spanish: **332 million**
English: **322 million**

Most body piercings on a woman (Elaine Davidson, UK): **720**
year: **2001**

Loudest burp (Paul Hunn, UK): **107.1 decibels**
year: **2008**

Highest altitude road in the world (in Khardungle Pass, Kashmir):
5,682 metres

Think of how stupid the average person is, and realize half of them are stupider than that.

George Carlin, US stand-up comedian, social critic, actor and author

***

Length of the longest female dreadlocks (Asha Mandela, USA):
2.59 metres
year: **2008**

Distance of the surface of the Dead Sea below mean sea level (the lowest point on the Earth):
417 metres
average full days of sunshine it receives each year:
330

Number of small biographies listed in 2009 Who's Who in Australia: **13,378**
proportion who are women:
21 per cent
who are academics and educators:
nearly 25 per cent

Area of the South China Sea:
about 3.5 million square kilometres
depth (in the north-eastern section):
up to 6,000 metres

Height of the Angel Falls in Venezuela (the world's highest):
979 metres

Area of Lake Superior in North America (the world's largest):
82,414 square kilometres

Estimated greatest depth of the:
Pacific Ocean: **11,034 metres**
Atlantic Ocean: **9,460 metres**

Area of the South China Sea:
2,974,600 square kilometres

Height of the CN Tower in Toronto, Canada (world's tallest free standing structure):
553.3 metres

Number of deserts in Australia: **10**

Estimated proportion of the world's population that live on less than $1.20 per day:
20 per cent

Number of languages spoken in Papua New Guinea (the most of any country): **about 720**

Number of people per square kilometre in Monaco (world's most densely populated independent country): **18,285**
Mongolia (the least): **1.7**

Number of countries that border on Turkey:
8 (Greece, Bulgaria, Armenia, Azerbaijan, Georgia, Iran, Iraq and Syria)

When you're right, no one remembers.
When you're wrong, no one forgets.

Anonymous

***

Total area of Austria:
83,870 square kilometres
population:
about 8.2 million
proportion who are Roman Catholic:
about 74 per cent

Number of continents that border on the Atlantic Ocean:
3 (America, Africa, Europe)

Maximum depth of the world's deepest lake (Lake Baikal, Russia):
about 1,637 metres
proportion of the world's unfrozen freshwater reserves it contains: **about 20 per cent**

Number of people who work in the Vatican:
about 3,000
number of clergy and nuns in Vatican City:
about 1,000

Number of time zones in Russia: **11**

Number of countries that border India:
6 (Bangladesh, Bhutan, Burma, China, Nepal, Pakistan)

Number of official languages of Equatorial Guinea:
2 (French and Spanish)

Number of visitors in 2008 to the:
Louvre, Paris, (the largest for a museum): **8.50 million**
British Museum, London: **5.93 million**
National Gallery of Art, Washington: **4.96 million**
Tate Modern, London: **4.95 million**

Height of Mount Ossa (highest mountain in Tasmania):
1,614 metres

Length of the Volga (longest river in Europe):
3,692 kilometres

Year Nelson Mandela was released from prison in South Africa:
1990
time he served: **27 years**

Number of men who in 2008: **6,906**
women: **12,894**

Most tap dancing taps in one minute (Tony Adams, New Zealand): **602**
year: **2008**

Average number of fog days in Canberra each year (the largest of any capital city): **47**

Join the Army. Visit exotic places. Meet strange people. Then kill them.

Private Joker (in the movie Full Metal Jacket}

***

Number of countries that border Kenya:
5 (Ethiopia, Somalia, Uganda, Tanzania, Sudan)

Number of migrants arriving from China to Australia between July and October 2009: **6,350**
July to October 2008: **5,540**

Number of migrants arriving from Britain to Australia between July and October 2009: **5,800**
July to October 2008: **8,110**

Average length of time that surveyed Britons sleep each night:
6 hours 17 minutes

Proportion of students in Australian universities who drop out in their first year of study:
20 per cent

Average life expectancy for boys born in Australia in 2006–8:
79.2 years
born in 1901–1910: **55.2 years**

Average life expectancy for girls born in Australia in 2006–8:
83.7 years
born in 1901–1910: **58.8 years**

Length of the Andes in South America (world's longest mountain range):
over 7,000 kilometres
average height: **4,000 metres**

Number of visits to national parks in the US:
in 2008: **274.9 million**
in 2003: **266.1 million**
in 1999: **287.1 million**

Estimated number of people aged over 85 in Australia in 2050:
1.3 million
number in 2009: **400,000**

Proportion of surveyed men in Australian who do not wash their hands after using a public toilet: **29 per cent**
of women: **8 per cent**

Proportion of Australians who selected 'no religion' in the 2006 census:
18.7 per cent
of New Zealanders who did the same thing in their 2006 Census:

32.2 per cent

Approximate size of the world's population in October 2010:
6.86 billion
proportion who live either in India or China:
37 per cent

I couldn't repair your brakes, so I made your horn louder.

Steven Wright, US actor and writer

***

Number of countries that border on Afghanistan:
6 (China, Iran, Pakistan, Tajikistan, Turkmenistan, Uzbekistan)

Number of students from India studying at US colleges and universities in 2008–9:
103,260
from mainland China: **98,235**
from South Korea: **75,065**

Population of China: **1.3 billion**
proportion who live in cities in 2010: **45 per cent**
in 1979: **17 per cent**

Estimated number of centenarians in Australia in 2020:
about 12,000

Number of arrivals to Australia in 2009:
12.4 million
number of departures:
12.3 million

ENTERTAINMENT

My father hated radio and could not wait for television to be invented so he could hate that too.

Peter De Vries, US editor and novelist

Television is to news what bumper stickers are to philosophy.

Richard Nixon, former US President

***

Gaming revenue by casinos on the Las Vegas strip in February 2010 the largest single-month increase since November 1999):
$US568 million
in February 2009:
$US427.4 million

Reported number of television advertisements seen annually by the average teenager in the USA:
about 40,000

Proportion of surveyed Americans who knew that Superman came from the fictional planet Krypton:
60 percent
that Mercury was the closest planet to the Sun:
37 per cent

DVD sales and rentals in the US in 2009:
$US16.4 billion

Proportion of novels sold in the USA that are romance novels:
34 per cent

Times of the day that Frasier Crane's radio show was usually broadcast:
2.00pm–5.00pm
station call sign: **KACL 780AM**

Reported sales of Bratz dolls sold worldwide since 2001:
over $US3 billion

Homer Simpson's social security number:
568-47-008

Estimated number of television sets in:
China: **about 400 million**
the USA: **about 220 million**
Japan: **about 87 million**
India: **about 63 million**

Number of films acted in by John Wayne before he won an Oscar for the movie True Grit: **138**
year: **1969**

Total number of words in the first six Harry Potter novels:
885,943

Proportion of surveyed Americans who feel that Hannibal Lecter is Hollywood's scariest villain (the highest):
23 per cent

Freddy Kruger: **17 per cent**
Michael Myers: **16 per cent**

Average amount of television watched by Americans each day:
4 hours 30 minutes

Number of squares normally found in a Snakes and Ladders Board: **100**

Dad, tell me again how when you were my age you had to walk all the way across the room to change a channel.

Anonymous

***

Number of DVDs sold in Australia in 2007: **82 million**
amount spent on them: **$1.46 billion**

Estimated number of copies of Michael Jackson's *Thriller* has sold worldwide:
over 100 million

Proportion of surveyed Australians who always or occasionally watch a code of football on television:
58 per cent

Proportion of complaints about TV advertising that involve sex or nudity:
30 per cent

Proportion of surveyed Americans who could identify any two of Snow White's seven dwarves:
77 per cent
any two US Supreme Court justices:
24 per cent

Proportion of Americans who always or often watch television while eating:
40 per cent

Amount paid at a Bonham's auction in London for the dress worn by Judy Garland in the 1939 movie *The Wizard of OZ:*
£140,000
year: **2005**

Proportion of surveyed Americans who say they learn about health issues from television:
88 per cent

Estimated proportion of two-year-olds in the US who watch at least an average of 90 minutes of television per day:
90 per cent

Mainland China's box office receipts in 2009:
$US890 million
in 2007: **$US455 million**

Estimated proportion of Chinese people that watch English Premier League matches on television compared to British people:
10 to 1

World Box Office revenue in 2007:
$US26.7 billion

amount of this from the US and Canada:
$US9.6 billion

Number of squares on a scrabble board: **225**

On cable TV they have a weather channel – 24 hours of weather. We had something like that where I grew up. We called it a window.

Dan Spencer

***

Number of Australians who watched at least part of the opening ceremony of the 2008 Beijing Olympics live on television:
about 8 million

Number of the house in Evergreen Terrace, Springfield, where the Simpsons live: **742**

Number of stars that currently surround the Paramount Picture logo: **22**

Cost of a 30-second TV commercial during the 2010 Superbowl in February:
about **$US2.8 million**
average number of viewers:
106.5 million

Average amount of time each week Australians spent listening to commercial radio during 2008:
17 hours 2 minutes

Amount spent on video games in the US in 2007:

$US17.94 billion
on video games hardware:
$US7.04 billion

Amount paid at auction in March 2008:
for a poster of the 1933 film "King Kong":
$US345,000
for Michael Keaton's complete Batman costume from "Batman Returns": **$US103,500**
for Julie Andrews' "Mary Poppins" coat: **$US51,750**

Amount paid in January 2007 on an overdue library book (Prince of Egypt) by Robert Nuranen to a Michigan library:
$US171.32
length of time it was borrowed: **47 years**

Amount paid at an internet auction in March 2009 for the first comic book featuring Superman:
$US317,200
year it was published: **1938**
original cost: **$US1**

Proportion of surveyed Australian women who said they cry at sad songs: **77 per cent**
men: **52 per cent**

Proportion of surveyed Australians who said they have to listen to music while cooking:
25 per cent

Proportion of surveyed Australians aged 18–24 who said they wouldn't leave the house without their personal stereo:
28 per cent

Number of black pieces in a game of draughts: **12**

I find television very educating. Every time somebody turns on the set, I go into the other room and read a book.

Groucho Marx, US comedian and actor

***

Average number of words each day people reportedly receive through the internet, email, television and other media:
100,500

Number of people treated in hospitals for fireworks injuries each year in the US:
about 8,000
proportion of these from backyard fireworks:
60 per cent

Number of strings on a normal classical violin: **4**

Number of tiles in a mahjong set: **144**

Year that actor Christopher Reeve died: **2004**
his age: **52**

Year that Luciano Pavarotti died: **2007**
his age: **71**

Number of marriages of actress Elizabeth Taylor: **8**

actor Mickey Rooney: **8**

Year of the first screen kiss (May Irwin and John Rice in 'The Widow Jones'):
1896

Most kisses by an actor in one movie (by John Barrymore in 'Don Juan'): **191**
year: **1926**

Number of different buses used in the movie 'Speed': **12**

Number of US Presidents to wear a Nazi uniform in a movie:
1 (Ronald Reagan in 'Desperate Journey')
year: **1942**

Reported number of illegal movie downloads each month in Singapore:
350,000

Proportion of surveyed Australians who associate times in their lives with certain songs:
83 per cent

It's a scientific fact. For every year a person lives in Hollywood, they lose two points off their IQ.

Truman Capote, US author

***

Average time that US males spend each month talking on mobile phones:
372 minutes
females: **275 minutes**

Proportion of surveyed Australians who said they couldn't drive more than two kilometres without the stereo playing in their car:
34 per cent

Number of Oscars won by Walt Disney: **32**

Year Canadian actor John Candy was born: **1950**
year he died: **1994**

Proportion of Australians who gamble regularly:
15 per cent
amount Australians lose gambling each year:
$18 billion
amount of this lost on poker machines:
$12 billion

Number of tiles in a standard double-six domino set: **28**

Number of aces in a 'dead man's hand': **2**
number of eights: **2**

Number of hours per week that the average American watches television: **35 hours**
timeshift (recorded) television: **2 hours**
internet: **4 hours**
online video: **22 minutes**

Average number of television sets in Australian households: **3**
average time Australian males spend each week watching television: **12 hours**

Number of types of woodwind instruments played in a classical orchestra:
4 (flute, oboe, clarinet and bassoon)

Proportion of Australian households that have at least one television: **99.7 per cent**
that had a digital television as at January 2010:
61 per cent

Year the first episode of *EastEnder* s was broadcast by the BBC:
1985
number of episodes so far:

over 4,000

Proportion of viewing share in Australia between 6.00pm–midnight that is free to air television:
83.4 per cent
pay television:
16.6 per cent

When an actor marries an actress they both fight for the mirror.

Burt Reynolds, US actor

***

Average number of television sets per household in the USA:
2.9
proportion of households with 3 or more sets:
55 per cent

Estimated amount that Australians spent in 2010 on online casino games, poker and bingo:
$968 million
proportion of this spent on poker alone:
33 per cent

Average number of television viewers in Australia who watched Roger Federer beat Andy Murray in the 2010 Australian Open Tennis final:
2,352,000

Most swords swallowed simultaneously (Chayne Hultgren, age 21, in Sydney): **18**
length of the blades: **72 centimetres**
year: **2010**

Year the f-word first appeared in a movie (I'll Never Forget What's 'isname): **1968**

Date the first posthumous Oscar was awarded (to Sidney Howard, best screenplay):
29th February 1940

SCIENCE, SPACE AND MATHEMATICS

Women have a passion for mathematics. They divide their age in half, double the price of their clothes, and always add at least five years to the age of their best friend.

Marcel Achard, French playwright and screenwriter

In the school I went to, they asked a kid to prove the law of gravity and he threw the teacher out of the window.

Rodney Dangerfield, US comedian and actor (Back to School, 1986)

***

Estimated volume of gold ever mined by 2010:
161,000 tonnes
volume: **8,333 cubic metres**

Number of decibels of:
the threshold of aural pain: **130**
a hair dryer: **70**
a dishwasher: **60**
a refrigerator: **50**

Year that Australia launched its first satellite (from Woomera, South Australia):
1967

Atomic number of oxygen: **8**
atomic weight: **15.9994**

Number of permutations of Rubik's cube:
43,252,003,274,489,856,000
number of correct ones: **1**

Estimated proportion of the Earth's land surface that is that is permanently frozen:
15–20 per cent

Proportion of the Earth's chemical composition by mass that is iron:
34.6 per cent

Estimated distance to the closest black hole from Earth (V 4641 Sgr):
1,600 light years

Number of the first 23 NASA astronauts who were the first born in their family: **21**

Successful space launches by the Soviet Union/Russia between 1957–2005: **2,746**
USA: **1,305**
European Space Agency: **160**
China: **88**
Japan: **62**

Reported amount of funding that NASA received:
in 2010: **$US17.2 billion**
in 2008: **about $US18.7 billion**

Number of operations per second the Earth Simulator in Japan is capable of:
35.8 million

Proportion of surveyed Americans who believed that intelligent aliens have visited Earth:
45 per cent

Number of points in Moh's scale of mineral hardness:
10

Britain has invented a new missile. It's called the civil servant – it doesn't work and it can't be fired.

Walter Walker, British General (quoted in newspapers, 1981)

***

Time after lift-off that the space shuttle Challenger exploded:
73 seconds
year: **1986**

Number of sides on a tetragon: **4**
trapezoid: **4**

The Atomic Number of hydrogen: **1**
zinc: **30**
krypton: **36**

Weight of the Hubble Space Telescope:
about 11,110 kilograms
year it was launched: **1990**

Minimum wind strength for a Category 5 hurricane:
249 kilometres per hour

Proportion of energy used in the USA in 2005 that came from fossil fuels:
about 86 per cent

Number of ergs in a joule:
10,000,000

Depth of the continental lithosphere (the outermost layer of the Earth's crust):
about 150 kilometres

Number of light years in a parsec: **3,258**
number of kilometres: **30,823,286,000,000**

Number of zeros in a googol: **100**

Number of actively broadcasting satellites orbiting the earth in the GPS constellation: **31**
altitude of their orbit:
about 20,278 kilometres

Number of grams in a hectogram: **100**

Estimated number of stars in the Milky Way:
200 billion–700 billion

Estimated span of the Milky Way:
about 100,000 light years

Height of the stratosphere above the Earth's surface:
10–31 kilometres

Number of different ways six 2x4 Lego bricks of the same colour can be put together:

915,103,765

In any collection of data, the figure most obviously correct, beyond all need of checking, is the mistake.

Finagle's Third Law

***

Length of a Mean Sidereal Day (the time needed for the Earth to complete one rotation about its axis):
23 hours 56 minutes 4.09 seconds

Number of dots in the Morse Code for the letter B:
3
dashes: **1**

Average temperature on Jupiter:
about −153 degrees Centigrade

Number of litres of rain that has fallen when one millimetre is recorded over one hectare:
10,000

Coldest temperature ever recorded in the Arctic (in Siberia in 1885 and 1933):
−68 degrees centigrade

Number of cloud drops required to make a typical rain drop:
1,000,000,000

to make a small drizzle drop:
1,000,000

Air pressure per square metre at ground level:
10 tonnes

Proportion of surveyed Australians who believe that:
aliens have visited Earth: **67 per cent**
aliens exist: **80 per cent**
authorities are covering up evidence of aliens:
75 per cent

Number of Earth years it takes Pluto to make one orbit around the Sun: **248**

Number of faces on an icosahedron: **20**

Average distance from the Sun to the Earth:
149,597,871 kilometres
to Neptune: **4.495 billion kilometres**

Average distance from the centre of the Earth to the centre of the Moon:
384,403 kilometres
year that a person last set foot on the moon: **1972**

Surface area of the Earth:
about 509,600,000 square kilometres
proportion that is water:
about 70.9 per cent

Number of Australians who measure rainfall each day as part of the Bureau of Meteorology’s official network of observations:
about 7,000

Enough research will tend to support your theory.

Anonymous

***

Number of megawatts in a terawatt:
1,000,000
in a gigawatt: **1,000**

Range of surface temperatures on the planet Mercury:
−110 degrees C to +450 degrees C
average: **180 degrees C**

Average temperature on Mars: **−63 degrees C**
minimum: **−140 degrees C**
maximum: **20 degrees C**

Number of sides on a heptagon: **7**
The smallest abundant number: **12**
perfect number: **6**

Proportion of Australian households using solar energy in 2008:
8 per cent
in 2002: **5 per cent**

Greatest recorded rainfall in one year (Mawsynram, Meghalaya, India):
11,873 millimetres

in 24 hours (Cilaos, Reunion, Indian Ocean, 1952):
1,870 millimetres

Farthest distance from the Earth reached by humans (on Apollo 13):
400,171 kilometres
year: **1970**

Oldest man in space (John Glenn, Space Shuttle Discovery):
77 years
year: **1998**
length of his mission: **9 days**

Mathematics may be defined as the subject in which we never know what we are talking about, nor whether what we are saying is true.

Bertrand Russell, British philosopher, mathematician, historian, socialist and pacifist

***

Number of years between 1998 and 2026 that have three Friday 13th dates:
5 (1998, 2009, 2012, 2015, 2026)

Year John Glenn became the first astronaut to orbit the earth:
1962

Speed at which a NASA spacecraft crashed into the moon in October 2009 to probe for water:
9,000 kilometres per hour

First number of the Fibonacci sequence: **0**
second number: **1**
third number: **1**

Atomic number of:
plutonium: **94**
oxygen: **8**
carbon: **6**

Date that Halley's comet will return to Earth:
28 July 2061

Number of stars a human eye can see on the clearest night:
up to 2,500

Number of years since the Neanderthals became extinct:
about 30,000

Lowest temperature recorded in the USA (at Prospect Creek, Alaska):
−62 degrees C
when: **23rd January 1971**

Number of times per second a quartz crystal in a wristwatch vibrates:
32,768

Estimated proportion of television static that derives from the Big Bang: **1 per cent**

Number of thunderstorms happening at any one time in the world:
about 1,800

Most distant object detected in the universe (star GRB 090423):
13.035 billion light years

year it was found (by NASA's 'Swift' satellite): **2009**

Year that Apollo 13 was launched: **1970**

FOOD AND BEVERAGE

I told my wife that a husband is like a fine wine; he gets better with age. The next day, she locked me in the cellar.

Anonymous

Actually, it only takes one drink to get me loaded. Trouble is, I can't remember if it's the thirteenth or fourteenth.

George Burns, US comedian, actor and writer

***

Production of red/rosé wine in Australia in 2008–9:
623.1 million litres
white wine: **537.3 million litres**

Estimated number of Australians affected by food poisoning each year:
about 5.4 million

Estimated average weight of vegetables (including potatoes) consumed each year by
Australians: **162 kilograms**
Americans: **189 kilograms**

Proportion of surveyed Australians who say they have gone to work under the influence of alcohol at least once in the previous twelve months:
7 per cent

Proportion of tea drunk in America that is iced:
85 per cent

Reported weight of food additives that people in industrialised countries eat each year:
6–7 kilograms

Number of beer bottles opened in one minute by a 30-year-old man in China by using his teeth: **41**

Average number of cups of tea drunk by Britons in their lifetime: **74,802**
apples eaten: **5,272**

Proportion of thefts from shops in Italy that is parmesan cheese (making it the most stolen item):
10 per cent

I don't drink these days. I am allergic to alcohol and narcotics. I break out in handcuffs.

Robert Downey Junior, US actor and musician

***

Proportion of surveyed Americans who say they like the taste of fast food too much to give it up:
44 per cent
Canadians:
37 per cent

Estimated proportion of all evening meals in Australia that are:
home cooked: **78 per cent**
takeaways: **10 per cent**
dining out: **9 per cent**
home deliveries: **3 per cent**

Reported number of people in India who rely on animal waste and firewood for cooking:
about 700 million

Proportion of surveyed Australian households that eat red meat at least once a week:
88 per cent
once a month: **96 per cent**

Estimated value of food thrown out by Australians each year:
$5 billion
its weight: **about 3 million tonnes**

Number of cups of tea drank each day by Australians:
22 million

Proportion of surveyed Australians who say that Chinese or Italian is their favourite type of cuisine:
30 per cent
traditional Australian or English cuisine:
13 per cent

Proportion of surveyed Australian parents who are not personally interested in cooking:
67 per cent

Proportion of a peanut that is cholesterol: **0**

Number of litres of wine in a magnum: **1.5**

Weight of the world's heaviest cake (Bucharest, Romania, inscribed with 'Happy New Year 2009'):
281 kilograms

World's longest smoked sausage (Bucharest, Romania):
392 metres
weight: **150 kilograms**

year: **2008**

Estimated number of hot dogs eaten each year in the US on their Independence Day weekend:
150 million
between Memorial day and Labor Day (hot dog season):
7 billion

Proportion of people who choke on food that reportedly do so on the first bite:
50 per cent

I envy people who drink – at least they know what to blame everything on.

Oscar Levant, composer, pianist, author, comedian and actor

***

Proportion of Australian beef production that is exported:
65 per cent
the value of these exports:
about $5 billion

Of those respondents who drank alcohol every day, proportion who had taken conventional medicine in the previous
24 hours: **84 per cent**
complementary medicine: **55 per cent**
both: **39 per cent**

Amount of wine exported from Australia in the year ending September 2009:
758 million litres
its value: **$A2.4 billion**

Proportion of the world's population who intentionally eat insects: **8 per cent**
the weight of insects that the average person eats each year:

about 1 kilogram

Proportion of surveyed Americans who eat a beef burger at least once a week: **51 per cent**
who have tomato sauce on it: **47 per cent**
mustard: **22 per cent**
mayonnaise: **20 per cent**

Proportion of surveyed Americans who say that chocolate chip is their favourite cookie flavour: **53 per cent**
peanut butter: **16 per cent**
oatmeal: **15 per cent**
sugar/shortbread: **11 per cent**
other: **5 per cent**

Proportion of surveyed Americans who would like to see detailed nutritional information on restaurant menus:
38 per cent

Proportion of surveyed Americans who say they never drink alcohol: **25 per cent**
proportion of these that are women: **29 per cent**
who are younger than 64: **72 per cent**

Proportion of surveyed Australians who consider the meat pie to be their national dish: **37 per cent**
roast lamb: **28 per cent**
lamington: **12 per cent**

pavlova: **11 per cent**
pasta bake: **1 per cent**

Weight of the world's heaviest gingerbread man (unveiled in Oslo, Norway):
651 kilograms
year: **2009**

Proportion of a cucumber that is water:
about 95 per cent

The laziest man I ever met put popcorn in his pancakes so they would turn over by themselves.

W.C. Fields, US comedian and actor

***

Weight of the durian fruit (also known as 'king of fruit'):
up to 4.5 kilograms
height of the durian tree:
25–50 metres

Proportion of Australians who believe that getting drunk occasionally is part of being Australian:
50 per cent
who feel pressured to drink alcohol in a workplace situation:
17 per cent
who know someone with an alcohol problem:
33 per cent

Proportion of the world's caviar that comes from the Caspian Sea: **90 per cent**
proportion of caviar that is protein:
25 per cent
fat: **19 per cent**

Proportion of Australians who feel the advertising of fast food is unacceptable: **51 per cent**

of alcohol: **53 per cent**
of gambling: **79 per cent**

Year that McDonalds opened their first fast-food restaurant in mainland China (in Shenzhen): **1990**

Proportion of surveyed Australian women aged 16–49 who said that eating chocolate makes them feel remorseful:
50 per cent
that eating berries makes them feel good:
84 per cent

Number of lunch boxes (including duplicates) on display at the Lunch Box Museum in Columbus, Georgia, USA:
over 3,500

Proportion of alcohol in 175 proof in Britain:
100 per cent
100 proof: **57.1 per cent**
80 proof: **45.7 per cent**
50 proof: **28.6 per cent**

Average amount of pasta eaten by an Italian each year:
about 54 kilograms

Weight of cheese reported stolen from a refrigerated 14.6 metre-long semi-trailer in Pennsylvania in May 2007:
10,433 kilograms

Reported weight of plastic drink bottles used in Australia each year:
about 118,000 tonnes
proportion that are recycled:
35 per cent

Proportion of surveyed Australians who confessed to eating junk food when depressed:
40 per cent

My old man, he was very strict. He allowed no drinking in the house. I had two brothers who died of thirst.

Rodney Dangerfield, US comedian and actor

***

Estimated amount of bottled water consumed by Australians in 2008:
about 300 million litres
amount spent on them:
about $465 million
proportion of water bottles that end up as landfill:
about 65 per cent

Proportion of surveyed men in Australia who had consumed alcohol in the previous week:
68 per cent
women: **51 per cent**
proportion of these that posed a low health risk:
79 per cent

Reported proportion of Australian adults who consume alcohol every day: **8 per cent**
are frequent users of ecstasy: **3 per cent**
cocaine: **1 per cent**
heroin: **1 per cent**

Length of the longest hot dog (made in Mexico):

114.32 metres
year: **2008**

Proportion of surveyed Australian children who take chips to school: **46 per cent**
who have fruit in their lunch box each day: **59 per cent**
sweets: **20 per cent**

Proportion of surveyed Australian mothers who do most of the family cooking: **78 per cent**
fathers: **3 per cent**
shared: **19 per cent**

Proportion of surveyed Australians who eat five vegetable serves daily: **20 per cent**
two pieces of fruit: **50 per cent**

Proportion of surveyed university students in Australia who drink a hazardous/harmful amount of alcohol: **34 per cent**
ages of the heaviest drinkers: **18–19 years**

Age range in Australia of the heaviest drinkers (most likely male, single, earning a good income): **18–35 years**

Proportion of surveyed US companies serving alcohol at their end-of-year holiday party in 2008:
71 per cent
in 2000: **90 per cent**

The worst gift is a fruitcake. There is only one fruitcake in the entire world, and people keep sending it to each other.

Johnny Carson, former US talk show host

***

Proportion of surveyed Australian mothers who say their husband is a better cook than them:
10 per cent

Average consumption of beer per person in Australia for the year ending 30 June 2009:
4.49 litres
in 1979: **6.40 litres**

Weight of the world's heaviest beef burger (at Randwick, Sydney): **95.5 kilograms**
number of eggs used: **120**
year: **2010**

Proportion of Italians who regularly sit down to a proper Sunday lunch with relatives each week:
85 per cent
who buy food made in Italy if possible:
90 per cent

Number of 7–Eleven stores in Japan (the most of any country) in mid-2010:

12,349
number of these in Tokyo: **1,577**

CHILDREN AND TEENAGERS

When I was a boy of fourteen, my father was so ignorant I could hardly stand to have the man around. But when I got to be twenty-one, I was astonished at how much the old man had learned in seven years.

Josh Billings, American humourist

Until I was 13, I thought my name was 'Shut up'.

Joe Namath, US footballer

***

Proportion of Australian households that have at least one child aged under 15:
29 per cent

Average number of US Presidents an 8–12-year old American child can name: **4.8**
alcoholic beverages: **5.2**

Proportion of surveyed Australians aged 12–15 years who smoke: **7 per cent**
aged 16 years: **17 per cent**

Proportion of surveyed Australian girls aged 15–17 who say that their mother is the most powerful influence in their perception of beauty:
1 in 3

Proportion of surveyed Australian children aged 7–14 who:
frequently use the internet: **89 per cent**
have a computer or video gaming system: **91 per cent**
access to an iPod or MP3 player: **63 per cent**

sent text messages while watching television: **36 per cent**

Proportion of surveyed Australian children aged 5–10 years who get a weekly allowance:
40 per cent
average proportion that boys receive more than girls:
25 per cent

Proportion of surveyed children aged 7–16 in Britain who believe that not having a mobile phone is a sign of poverty:
20 per cent
not having books at home:
20 per cent
not having a proper school uniform:
40 per cent

Proportion of surveyed boys in US high schools who play cards for money at least once a week:
11 per cent

Percentage of US-born youths aged 16–19 in the USA who are not enrolled in an educational institution or working:
7.6 per cent
of non-US citizens: **13.3 per cent**

Proportion of surveyed Australian males aged 14–19 who have used an illicit drug in the past year:

15.6 per cent
females: **17.7 per cent**

Proportion of surveyed Australian children aged 5–9 who go home alone after school:
3 per cent
aged 10–15: **20 per cent**

Estimated number of homeless people in Australia:
about 105,000
number of these aged 12–18: **about 18,000**
proportion of homeless who are married: **55 per cent**
couples without children: **20 per cent**

My Father had a profound influence on me, he was a lunatic.

Spike Milligan, Irish comedian and writer

***

Proportion of surveyed 12–25 year-olds in Australia who take prescription drugs recreationally:
23.5 per cent
of those, 12–14 year-olds (the highest group):
24.9 per cent

Proportion of surveyed Australian males aged 14–19 who drink alcohol every day:
1.4 per cent
females: **0.5 per cent**

Proportion of children aged 8–11 years in Australia who own a mobile phone: **18 per cent**
12–14 years: **75 per cent**
15 years and over: **90 per cent**

Proportion of surveyed 15–17 year old boys in Australia who had consumed alcohol in the previous week:
24 per cent
girls: **16 per cent**

Number of Australian families with children aged under 15 in 2008: **3.2 million**
number of these in New South Wales: **1.3 million**
in Victoria: **950,400**

Proportion of Australian parents who feel it is acceptable to let their children drink before the legal age:
24 per cent
who are happy to serve alcohol to children aged under 16:
11 per cent

Proportion of surveyed Australian parents who think their daughters are growing up too fast:
86 per cent

Proportion of surveyed 7–14 year olds in Australia who receive no pocket money: **32 per cent**
average amount boys receive per week: **$7.60**
girls: **$6.80**
proportion more likely a boy is to have a girlfriend if they receive $5 pocket money or more per week:
80 per cent

Proportion of Australians aged 15–64 enrolled in a course of study as at May 2008:
19 per cent
proportion of these leading to a qualification:
96 per cent

Reported proportion of Australian children aged 24–27 months who are fully immunised:
about 92 per cent

Total number of hours spent in the classroom by children in Australia aged between 7 and 14:
about 8,000

Parents are the last people on earth who ought to have children.

Samuel Butler

***

Proportion of primary school teachers in Australia who are women:
80 per cent

Proportion of surveyed young Australians who say smoking marijuana is relaxing: **60 per cent**
who describe it as 'fun': **25 per cent**

Estimated number of teenagers in Australia each year who start smoking:
45,000

Estimated number of children worldwide each year who die before age 5:
about 9 million

Proportion of surveyed Australian males aged 14–19 who smoke every day:
6.0 per cent
females: **8.7 per cent**

Proportion of surveyed Americans adults who would allow their healthy child to participate in medical

research if the risk of harm was small: **25 per cent**
who would refuse: **38 per cent**
who are unsure: **37 per cent**

Proportion of surveyed Hong Kong fathers who did not play at all with their children in the previous week:
31 per cent
who did so for less than 15 minutes:
24 per cent

Proportion of surveyed Australian children aged 14–16 who do not eat enough fruit:
99 per cent
enough vegetables:
95 per cent
of 14–16 year old girls who are not getting enough calories:
80 per cent

Proportion of surveyed fathers in the UK who admit they enjoy going to work since it gives them a break from their children:
62 per cent
who said they secretly hoped their first born would be a boy:
67 per cent

Proportion of Australians who are Generation Y:

19 per cent
years of birth: **1980–1994**
their estimated proportion of the workforce in 2020:
42 per cent

Probability of a child being left-handed if:
both parents are right-handed: **9.5 per cent**
one parent is left-handed: **19.5 per cent**
both parents are left-handed: **26 per cent**

Experts say you should never hit your children in anger. When is a good time? When you're feeling festive?

Roseanne Barr, US comedienne

***

Proportion of surveyed Australian girls aged 11–18 who:
wish they weighed less: **60 per cent**
know someone with an eating disorder: **45 per cent**
would get cosmetic surgery if they could: **27 per cent**

Proportion of surveyed girls aged 10–18 who said they would feel better about themselves if they lost weight:
46 per cent
who said they were unsatisfied or not at all satisfied with their bodies: **39 per cent**

Maximum allowable working hours per week in a coal mine for boys aged 13–18 years in NSW in 1876:
50.5

Proportion of surveyed six-year-old British girls who said they would like to be thinner:
50 per cent

In the 2006 Australian Census, the proportion of women who had just one child:
13 per cent

Estimated proportion of male school pupils in Australia who are chronically sleep deprived:
42 per cent
female: **40 per cent**

Proportion of surveyed Australian teenagers who receive money from their parents for doing chores at home:
27 per cent

Proportion of surveyed British schoolchildren aged 9–15 who thought that Hitler:

was Germany's national football coach: **7 per cent**
invented gravity: **13.5 per cent**

HOUSE AND GARDEN

Christmas at my house is always at least six or seven times more pleasant than anywhere else. We start drinking early. And while everyone else is seeing only one Santa Claus, we'll be seeing six or seven.

W.C. Fields, US comedian

The best time to plant a tree is twenty years ago. The second best time is now.

Chinese proverb

***

Proportion of Australian households that rely on welfare payments:
13 per cent

Estimated proportion of Australian homes that have at least one empty bedroom:
77 per cent

Proportion of surveyed Australian households that use cold water in their washing machines:
74 per cent
in 1994: **61 per cent**

According to the 2006 Census, proportion of Australians living in rural areas: **12 per cent**
in 1911: **42 per cent**

Proportion of surveyed Australian households in 2008 that had a clothes dryer: **56 per cent**
heater: **77 per cent**
cooler: **67 per cent**

Average time men spent each week in housework:

in 2006: **18 hours 20 minutes**
in 1992: **16 hours 55 minutes**

Rate of growth of the fastest growing weed (Japanese Knotweed): **10 centimetres/day**
depth of roots: **2 metres**
height: **1–3 metres**

Maximum height of a sunflower: **8 metres**
diameter of the corolla: **80 centimetres**

Average weight of a coconut (fruit): **8 kilograms**
maximum weight: **15 kilograms**

Length of a petal on Jacob's Ladder Orchid:
25–30 centimetres

Height of the 'Giant Saguaro' cactus found in the USA and Mexico:
up to 16 metres

Average percentage by which the price of established houses in the capital cities in Australia rose between June 2002 and June 2008:
about 10% per year

Proportion of surveyed Australian males aged 18–24 who still live with their parents:
49 per cent
of females: **45 per cent**

Proportion of Australians who live by themselves:
10 per cent

Gardening requires lots of water-most of it in the form of perspiration.

Lou Erickson

***

Proportion of households in Australian with an internet connection in 2007:
64 per cent

Number of houses and home units sold in Sydney in 2007:
about 94,500
in Melbourne: **about 93,000**

According to a survey, the proportion of times women have the final say when it comes to buying a new home:
62 per cent
men: **8 per cent**
joint: **30 per cent**

Estimated proportion of time the average Australian woman spends:
looking in the mirror each week: **3.5 hours**
cooking: **8 hours**
doing laundry: **3 hours**

Reported number of US households that had difficulty in putting enough food on a table at some point during 2008:
17 million
as a percentage of all households:
14.6 per cent

Proportion of home sales in the US in 2008–9 that were to first home buyers: **47 per cent**
in 2007–8: **41 per cent**

Amount of garbage generated by Americans each day:
2.1 kilograms
proportion of this that goes to landfill: **54 per cent**
recycled: **33 per cent**
burned: **13 per cent**

Estimated proportion of the weight of a two year-old pillow that is comprised of dead skin and dust mites:
10 per cent

Estimated number of Britons who wash their bed sheets three times or less a year:
500,000
age of those with the dirtiest sheets:
under 25 years

Proportion of surveyed Australians who have a high level of trust in their neighbours:
33 per cent
in members of their immediate family:
90 per cent

Proportion of surveyed Australians who say there is nobody they could visit without an invitation:
12 per cent

Home is where you can say anything you like cause nobody listens to you anyway.

Anonymous

***

Estimated proportion of Australia's homeless who are aged 24 years or less:
46 per cent

Proportion of Australian households that are occupied by renters:
28.5 per cent

Number of Australians aged 15 and over who care for someone at home because of disability or old age:
about 2.5 million
proportion of carers themselves who have a disability:
35 per cent

Average household size in Australia:
in 2006: **2.6**
in 1911: **4.5**
estimated number in 2025: **2.2 or 2.3**

Average amount of time Australian men spend each day doing unenjoyable chores:
50 minutes
women: **44 minutes**

Proportion of mortgaged US households in December 2009 in which more is owed on their mortgage than the house is worth:
25 per cent

Number of households in Australia in 2006:
7.8 million
proportion of these that have only one person:
23.8 per cent

Proportion of surveyed Australian homes that have at least one room devoted to clutter:
88 per cent
proportion of people who are psychologically affected by it:
40 per cent

Don't wear perfume in the garden – unless you want to be pollinated by bees.

Anne Raver, US nature and gardening journalist

***

Highest selling price of a home in Australia (in Mosman Park, Perth): **$57.5 million**
month it was sold: **December 2009**

Average size (in square metres) of new homes in 2009:
in Australia: **214.6**
in the USA: **201.5**
in New Zealand: **196.2**
in Greece: **126**
in Italy: **81.5**
in the UK: **76.2**

TRANSPORTATION

Every time my wife takes the car there's trouble. The other day she came home there were a hundred dents in the car. She said she took a shortcut through a golf range.

Rodney Dangerfield, US comedian

Life is like stepping onto a boat which is about to sail out to sea and sink.

Shunryu Suzuki Roshi, Japanese-born Zen master

***

Number of baggage items sent to departing aircraft on a busy day at Sydney International Airport:
about 19,000
peak times: **6am–10am and 12 noon–8pm**

Estimated number of caravans and campervans in use in Australia: **330,000**
number of caravan and tourist parks:
over 2,700

Proportion of surveyed Australian male drivers who have been involved in a crash:
84 per cent
of women: **77 per cent**

Total number of airline passenger flights in the world each year:
about 2 billion
number of bags checked in: **300 million**
number that go astray: **30 million**
number that are never recovered: **200,000**

Number of vehicles stolen in Australia each year:

about 85,000
rate per 1,000 registered vehicles in NSW: **1.7**
in Victoria: **1.1**

Length of the world's largest (in gross tonnage) passenger vessel "Freedom of the Seas":
338.77 metres
height: **63.7 metres**
gross tonnage: **140,000 tonnes**

Number of passengers in 2009 that passed through:
Atlanta Hartsfield Airport (highest in the world): **88.0 million**
London Heathrow: **66.0 million**
Beijing: **65.3 million**
Chicago O'Hare: **64.4 million**
Tokyo Haneda: **61.9 million**
Paris: **57.9 million**
Los Angeles International: **56.5 million**

Average number of bicycles manufactured worldwide in one hour: **11,500**
cars: **4,700**

Proportion of surveyed 55–64 year old Australians who plan to travel the world within the next five years:
1 in 3

Proportion of surveyed Australian drivers who say they have been a perpetrator of road rage:
74 per cent
proportion of these that were rude hand gestures:
53 per cent
abusive shouting: **49 per cent**
other: **8 per cent**

For Sale: Parachute. Only used once, never opened, small stain.

Anonymous

***

Proportion of surveyed Australian males who admit to grooming themselves in the rear vision mirror while driving:
25 per cent
females: **33 per cent**
age group of the worst offenders: **25–34**

Proportion of surveyed 18–24 year-old Australian males who read or send text messages while driving:
79 per cent
25 years and over: **37 per cent**

Length of the world's largest and heaviest aircraft, the Antonov AN-225: **84 metres**
wingspan: **88.4 metres**
empty weight: **about 175,000 kilograms**

Length of the smallest jet aircraft (in Puerto Rico, USA):
3.7 metres
wingspan: **5.7 metres**

weight: **162 kilograms**
top speed: **483 kilometres per hour**
year completed: **2004**

Proportion of surveyed British drivers who, to help them stay awake while driving, have slapped themselves across the cheek:
3 per cent
wound down the windows: **75 per cent**

Number of new bicycles sold in Australia:
in 2007: **1,203,648**
in 2001: **774,938**

Number of new motorcycles sold in Australia:
in 2008: **129,966**
in 2006: **119,209**

Proportion of surveyed Australian drivers who are concerned about the effect that their car is having on the environment:
79 per cent

Proportion of surveyed Australians who say that everyone should know first aid:
83 per cent
who feel confident that they could render basic first aid to a car accident victim:
7 per cent

Number of persons per motor vehicle in Australia in 2007:
1.44
in 1997: **1.66**

Number of passengers that passed through Sydney Airport in 2009: **33.4 million**
world ranking: **28th**

Reported blood alcohol reading of a motorist stopped after a minor accident in Bulgaria in August 2008: **0.851**
level at which it can be fatal: **0.400**
amount of beer he reportedly drank: **20 litres**

Never run after a bus or a man. There will always be another one.

Old Irish expression

***

Estimated proportion of Australians who still use a car as their main means of transport:
75 per cent
walking or cycling: **5 per cent**

Proportion of surveyed male drivers aged 18–75 in New Zealand who say they exceed the speed limit most of the time:
15 per cent
of women: **9 per cent**

Proportion of surveyed Australians who admitted to driving when they were probably over the limit:
37 per cent
who have taken a different route home just to avoid being breathalysed:
15 per cent

Estimated proportion of vehicles driven in the US that have an engine check light showing indicating a problem:
10 per cent

proportion of these where the light has been showing for at least three months:
50 per cent

Proportion of surveyed male New Zealand drivers aged 18–75 who describe themselves as impatient:
25 per cent
female: **19 per cent**

Proportion of surveyed Australian drivers aged 65 and over who had been involved in an accident during the previous 12 months:
8.5 per cent
aged 18–54: **20 per cent**

Proportion of workplace deaths in Australia that are due to:
vehicle accidents: **46 per cent**
falling from a height: **12 per cent**
being hit by a falling object: **10 per cent**
being hit by a moving object: **10 per cent**
electricity: **7 per cent**
other: **15 per cent**

Proportion of car journeys in Australia that are less than 1 kilometre: **10 per cent**
less than 3 kilometres: **30 per cent**
less than 5 kilometres: **50 per cent**

Estimated time the average American living in an urban area spends each year stuck in traffic:
38 hours

My mom said the only reason men are alive is for lawn care and vehicle maintenance.

Tim Allen, US comedian and actor

***

Number of motor vehicles registered in Australia as at 31 March 2009:
15,674,436
average age of the vehicles:
9.9 years

Proportion of children in the US who walk or ride a bike to school: **16 per cent**
in 1969: **about 50 per cent**

Number of Australians who had a cruise holiday in 2008:
330,949

Proportion of cars in China that are privately owned:
59 per cent
in 1991: **16 per cent**

Reported number of bags that went missing from the world's airports in 2009: **25 million**
proportion of these misplaced during aircraft transfers:
52 per cent
not loaded onto the plane:

16 per cent

Year that a plane first landed on a ship (piloted by Eugene Ely on to the armoured cruiser USS Pennsylvania):
1911

Proportion of surveyed Americans who think airlines are the safest way to travel: **45 per cent**
cars: **39 per cent**
trains: **16 per cent**

Year the first hot air balloon was flown (in Paris by Pilatre de Rozier and Marquis d'Arlandes): **1783**
height to which it rose:
about 1,000 metres

Year of the first flight across the USA (by Calbraith Rogers):
1911
number of stops along the way: **69**
time taken: **49 days**

Year the first helicopter (Gyroplane No. 1) flew: **1907**
time of flight: **about 60 seconds**
height reached: **60 centimetres**

Land speed record for a steam-powered vehicle (Charles Burnett III, UK):
225 kilometres per hour

year: **2009**
time the previous record had stood: **103 years**

Bicycles are almost as good as guitars for meeting girls.

Bob Weir, US singer, songwriter and guitarist

***

Sail area (mainsail plus working jib) of *Mirabella V* (the largest single masted yacht ever built):
2,385 square metres
height of the mast above the waterline:
89 metres

Number of people killed in traffic accidents in mainland China in 2006 (the lowest in six years):
about 89,455
in India in 2007 (the world's highest): **114,580**

Proportion of surveyed Australian women who feel they are treated more poorly than men at car dealerships:
90 per cent
who feel they are largely ignored in car showrooms:
25 per cent

Average time it takes pedestrians in London to cover 18.3 metres: **12.17 seconds**
in Edinburgh: **13.29 seconds**
in Cardiff: **16.81 seconds**
in Dublin: **11.03 seconds**

in Singapore: **10.55 seconds**

Largest wingspan on a plane (nicknamed the 'Spruce Goose'):
97.54 metres
length: **66.65 metres**
number of times it flew: **1**
date it flew: **2 November 1947**

Number of fatalities worldwide from aviation accidents (from planes carrying more than six passengers):
in 2009: **1,103**
in 2007: **971**
in 2000: **1,582**

Proportion of surveyed Britons who nominated airport check-ins as their most hated queue: **65 per cent**
who use queues to chat up fellow travellers:
18 per cent

The only way of catching a train I have ever discovered is to miss the train before.

Gilbert K. Chesterton

***

Proportion of surveyed Australians who believe that parking in Australian CBDs should be:
free: **29 per cent**
$2 per hour: **42 per cent**
up to $5 per hour; **15 per cent**
$5 or more: **14 per cent**

Proportion of motor vehicle passengers in Australia in 2008 who regularly wore seatbelts:
83 per cent
in 1988: **45 per cent**

Length of *Oasis of the Seas,* the world's largest cruise ship:
360 metres
weight: **225,282 gross tons**
passenger decks: **16**
crew: **2,165**
estimated cost to build: **$US1.4 billion**

Distance travelled in its first year of 1928 by the Australian Flying Doctor Service:
32,000 kilometres

number of flights: **50**

Road toll in 2006:
in India: **106,725**
in China: **89,455**
in the USA: **42,642**
in Brazil: **35,155**
in Russia: **33,308**
in Iran: **22,918**
in Australia (in 2009–10): **1,424**

Number of people worldwide each year who die in traffic accidents: **about 1.2 million**
proportion of these who are pedestrians, cyclists or motorbike riders: **50 per cent**

Number of jets worldwide that crashed:
in 2008: **19**
in 2007: **17**
in 1999: **24**

Proportion of surveyed Australian drivers who admit they have fallen asleep at the wheel:
25 per cent
of those aged 18-24: **38 per cent**

Number of US pilots who tested above 0.04 (their maximum legal blood alcohol level) between 1997 and 2008:
136

When the plane you are on is late, the plane you want to transfer to is on time.

Anonymous

***

Proportion of surveyed airline passengers who would prefer to go through a body scan machine in preference to a manual pat down: **90 per cent**
who would decline a body scan: **4 per cent**

Proportion of surveyed Australians who feel that price, quality and equipment are the most important attributes of a car:
34 per cent
how it drives: **33 per cent**
engine performance and efficiency: **20 per cent**
safety: **9 per cent**
comfort and practicality: **4 per cent**

Estimated proportion of passengers on a flight of two and a half hours or longer who will have a cold within a week:
20 per cent

Proportion of surveyed Australian motorists who have yelled or sworn at another motorist:
60 per cent

Number of portable sat-nav systems purchased in Australia in 2009:
more than 1 million

OTHER STUFF

I refuse to answer that question on the grounds that I don't know the answer.

Douglas Adams, English writer and musician

A classic is something that everybody wants to have read and nobody wants to read.

Mark Twain, US author and humourist

***

Number of British monarchs since William the Conqueror:
40

Proportion of surveyed Australians who were worried about the future of the environment:
91 per cent

Number of holistic veterinarians in Australia in 2010:
over 140
vet acupuncturists:
over 250

Proportion of Australians who regularly recycle cardboard, glass and plastic containers at home:
89 per cent
world average:
69 per cent

Proportion of residents of Britain who were born overseas:
10 per cent
of New Zealand: **19.5 per cent**

of Australia: **22 per cent**

Height of Ben Nevis, the highest mountain in Britain:
1,344 metres

Population of Sydney in 1901: **496,990**
Melbourne: **501,580**
Brisbane: **120,650**
Adelaide: **162,200**
Perth: **70,700**
Hobart: **36,060**

Proportion of Australians aged 65 and over in June 2010:
13 per cent
projection for 2050: **23 per cent**

Number of lines in a rondeau: **13**

Reported number of British soldiers who died at the Battle of the Somme (France, 1916): **19,240**
number wounded: **38,230**

Number of letters and parcels posted in Australia during the festive season:
about 450 million
proportion of surveyed Australians who prefer to receive their Christmas cards by traditional means:
91 per cent

Weight of the largest pearl (found in Palawan, Philippines, 1934): **6.3 kilograms**
diameter: **14 centimetres**

Population of Iran in 2010: **about 72 million**
proportion aged under 30 years: **60 per cent**

A man's got to believe in something. I believe I'll have another drink.

W.C. Fields, US comedian and actor

***

Estimated number of people worldwide who do not have adequate sanitation:
2.6 billion
as a proportion of the world's population:
about 40 per cent

Proportion of Australia's population who are male:
49.24 per cent
female: **50.76 per cent**

Proportion of Australians who are aged under 15 years as at 30 June 2009:
19.1 per cent

Year the English defeated the French at the Battle of Agincourt:
1415
reported French superiority in troop numbers:
6 to 1

Number of British people each day who leave the UK to live abroad: **500**
proportion of these who go to Australia:

20 per cent

Estimated number of people worldwide who can speak Spanish:
about 350 million

Estimated number of mobile phone messages sent in the world in 2008: **4.1 trillion**
estimated number in 2007: **1.9 trillion**

Height of the tallest tower (free standing structure) in the Southern Hemisphere (The Sky Tower, Auckland):
328 metres
year it opened: **1997**

Estimated proportion of glass in Australia that is recycled:
36 per cent
plastic: **31 per cent**

Average distance walked by Britons each year:
about 320 kilometres

Population of Indonesia in July 2010:
about 243 million

Estimated proportion of Australians who have a degree, diploma or certificate qualifications:
60 per cent

Number of months with 31 days: **7**

When I was young I was called a rugged individualist. When I was in my fifties I was considered eccentric. Here I am doing and saying the same things I did then and I'm labelled senile.

George Burns (Just you and me Kid, 1979)

***

Number of Popes of the Catholic Church since 21 June 1963: **4**

Rate of melanoma per 100,000 people in Queensland (the highest in the world): **65.3**

Proportion of the world's daily oil supply that is used by the USA:
about 24 per cent

Proportion of surveyed Australians aged 16–85 with a mental disorder: **20 per cent**
proportion of men: **18 per cent**
of women: **22 per cent**
proportion affected by an anxiety disorder (the most common):
14 per cent

Proportion of Australians who have at least one sight problem:
51 per cent

of those aged over 45 years: **87 per cent**
over 75 years: **96 per cent**

Estimated proportion of water in Australia that is used for agriculture:
64 per cent

Number of finger holes in a standard penny whistle: **6**

Estimated proportion of the world's population who will live in cities by 2030: **60 per cent**
who did so in the 1950s: **30 per cent**

According to the 2006 Australian Census, proportion of women aged 20–30 in Australia who hold a Bachelor degree or higher:
28 per cent
of men: **21 per cent**

According to the 2006 Census, proportion of Australians who live in towns and centres within 50 kilometres of the coast:
75 per cent

Number of words in the Gettysburg address: **about 272**
number of different known manuscripts: **5**

Length of a vicennial period: **20 years**

Proportion of Australians who will suffer a panic attack some time in their life: **1 in 3**
age group most likely: **18 to 35**

According to the 2006 Census, proportion of Australians who hold some sort of religious belief: **79 per cent**
in Sydney: **84 per cent**
Melbourne: **78 per cent**
Adelaide: **73 per cent**

Women like silent men. They think they're listening

Marcel Achard, Quote, 4 November 1956

***

Number of people who left Australia permanently in 2008–9:
81,018
proportion who were born in New Zealand: **9.9 per cent**
Great Britain: **7.6 per cent**
China: **7.2 per cent**
Hong Kong: **2.7 per cent**

Number of permanent new settlers to Australia in 2008–9:
158,021

Number of regular volunteers in Australia in 2006:
3.1 million
proportion of women who do so:
22 per cent
men: **19 per cent**
age group most likely: **35–44 years**

Number of international students enrolled at universities, TAFEs and private training institutions in Australia:
about 250,000

proportion of these who are from China:
24 per cent
India: **9 per cent**
South Korea: **7 per cent**

Number of muses in Greek mythology: **9**
Number of apostles: **12**
sacraments: **7**

Age to which Abraham lived according to the Bible:
175 years

Estimated known species of plants, animals and other organisms on Earth: **about 1.8 million**
estimated number: **5 to 100 million**

Reported number of active troops in the armed forces of China:
about 2.25 million
the USA: **about 1.47 million**
India: **1.4 million**
Russia: **1.04 million**

Number of sheets in a quire of paper: **24 or 25**
in a bale: **5,000**

Estimated number of foreign missionaries of all Christian
denominations worldwide:
about 306,000

Proportion of the world's coarse wool that comes from New Zealand:
40 per cent

Number of countries in Polynesia that have a monarchy:
1 (Tonga)

Always go to other people's funerals, otherwise they won't come to yours.

Yogi Berra, former US baseball player and manager

***

Number of central universities in India in 2009: **20**
state universities: **215**
deemed universities: **100**
colleges: **about 16,000**

In 1982, the number of times a lion and lioness were observed to have sex during a 5 hour 30 minute time period: **23**
number of these initiated by the lioness: **22**

Number of checkers each player has in the game of Backgammon: **15**

Total number of dots on a pair of dice: **42**

Number of members of the Arctic Council: **8**

Proportion of the world's diamonds that are used to make gemstones:
20–25 per cent

Number of countries in Africa (the largest of any continent):
53

Further Reading

Some of the many references consulted in the writing of this volume are listed below. They provide a rich source of numerical and other information.

Australian Bureau of Statistics
(URL: www.abs.gov.au)

Book of Historical Records (2000)
Norris McWhirter, Virgin Publishing

Chronicles of the 20th Century (1990)
Chronicle Publications, Mt. Kisko, New York

Durex Annual Global Sex Survey
(URL: www.durex.com/uk/globalsexsurvey)

Encyclopaedia Britannica Online
(URL: www.britannica.com)

Guinness World Records (annually 1956–2010)
(URL: www.guinnessworldrecords.com)

Hot Sex: how to do it (1999)
Tracey Cox, Pub. Bantam

Macquarie Book of Events (1983)
ed. Bryce Weldon, Pub. Weldon

Number Crunch (since 2000) John Croucher, weekly column,
Good Weekend, Sydney Morning Herald, Fairfax

Number Crunch (since 2000) John Croucher, weekly column,
Good Weekend, Melbourne Age, Fairfax

The Big Book of Records (2003)
ed. Maureen Spurgeon, Pub. Brown Watson, England

The Book of Numbers (2000)
William Hartston, Metro Pub. Ltd., London

The Odd Essential Body (2005)
Stephen Juan, Pub. Harper Collins, Australia

The Odd Sex (2001)
Stephen Juan, Pub. Harper Collins, Australia

This Book of More Perfectly Useless Information (2005)
Mitchell Symons, Pub. Transworld, UK

Vital Statistics (2005)
Paul Grobman, Pub. Plume, USA

What are the Chances of That? (2004)
William Hartston, Metro Pub. Ltd., London

Books For ALL Kinds of Readers

At ReadHowYouWant we understand that one size does not fit all types of readers. Our innovative, patent pending technology allows us to design new formats to make reading easier and more enjoyable for you. This helps improve your speed of reading and your comprehension. Our EasyRead printed books have been optimized to improve word recognition, ease eye tracking by adjusting word and line spacing as well as minimizing hyphenation. Our EasyRead SuperLarge editions have been developed to make reading easier and more accessible for vision-impaired readers. We offer Braille and DAISY formats of our books and all popular E-Book formats.

Printed in Dunstable, United Kingdom

91336855R00252